WEEKLY GEOGRAPHY PRACTIC

Table of Contents

Introduction

The modern world has woven a web of interdependency. Each part of the world gives to and takes from the other parts. We gain a better sense of where and who we are today through a knowledge of geography. By giving students geographical information and allowing them to use their own experiences, we can help them to connect the familiar in their lives with the unfamiliar. They can then move more easily from the known features of locale and state to the more unknown quantities of region, country, and world. Students can begin with the area they know and then expand their knowledge base. By giving students a better knowledge of geography, we give them a better understanding of the world in which they live. Such knowledge will help students in their standardized testing and in their broader academic pursuits.

General standards in geography for this grade suggest that students should be able to perform a variety of skills. Students should be able to identify landforms, understand and use map components, use latitude and longitude, and employ the cardinal and intermediate directions. They should understand regional similarities and differences and be able to discuss the relationship between humans and the environment. This book adheres to these standards.

Organization

The book provides a collection of 36 weeks of geography practice activities that highlight a different theme of geography each day of the week. On the back of each weekly practice student page, there is a corresponding teacher page with answers and additional information.

The book is organized using a logical movement from the most specific and familiar geographical concept—community—to the most general and most unfamiliar—Earth as a part of the solar system. This organization allows the students to understand better the increasing rings of relationship in their world.

Use

Weekly Geography Practice is meant to supplement the social studies curriculum in the classroom. Map skills help students to improve their sense of location, place, and movement. This book is designed as a flexible tool for the teacher. The practice lessons may be used in a wide variety of settings and modes. The weekly lessons may be used in sequence, or you may prefer to use the lessons wherever they best fit into your curriculum plans. You may photocopy the entire page of activities and hand them out to students at the beginning of the week, cut the pages into strips and pass them out daily, maintain the sheets in a folder to use at a specific time each day, assign the activities as homework, or use them as overhead transparencies.

The weekly lesson pages may be used for any of the following activities:

* warm-ups for the social studies class,
* cooperative group activities,
* independent student-monitored practices at a learning station,
* make-up assignments,
* additional assignments for students with specific needs,
* extra-credit assignments,
* or as short transitional activities between periods of longer concentration.

The Five Themes of Geography

An effective approach to the study of geography can be organized around the five themes of geography. The weekly practice questions are divided among five themes, each of which is represented by a graphic icon to serve as a visual reminder of the theme.

 **LOCATION**

Where is it? Location tells where a place is located. It can include latitude and longitude, cardinal or intermediate directions, or even just the words "next to."

 **PLACE**

What does it look like? What are its surroundings? Place is the part of an area you see. It includes landforms, rivers, seats of government, or even buildings in a city.

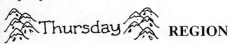 **MOVEMENT**

How do people, goods, and ideas move from one place to another? Movement looks at how and why things move. It looks at why people leave a country, how goods are moved, and how both affect the land or people.

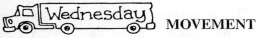 **REGION**

How are different places in an area alike? Region looks at the way an area is divided. A region has characteristics that are the same. Often areas are grouped because the people share a language, people do the same kind of work, or the climate or landforms are the same.

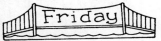 **INTERACTION**

How do people use or change the land? Interaction is how people use or change the environment. It considers where people have parks, build cities, set up factories, or run farms.

Assessments

Each of the five units in the book includes a pre-assessment test and a post-assessment test. As suggested, one assessment is intended to be completed before the unit of study is begun. This assessment gauges the students' basic knowledge of the unit's concept. The other assessment is intended to be completed when the unit of study is finished. This assessment gauges how well the students have learned the concepts presented in the unit.

Maps and Globes

Students gain a greater sense of place by knowing their relationship to other places. For this reason, maps and globes are important tools. This book includes 19 reproducible maps for use with the weekly practice sheets. Maps are also included for the assessment tests. One weekly practice sheet requires the use of a globe. In any case, various maps, especially of the United States and the world, are a handy addition to any classroom.

Additional Activities

Most of the weekly practice questions require short answers. Some questions are more substantial and require a longer answer. You might consider assigning these as homework. In addition, on page 3 is a series of additional activities tailored to each unit. These can be used as extra-credit assignments, research projects, or group work.

Notes

1. Each daily practice lesson should take no more than five to ten minutes for an average student.
2. Students may complete the lessons on their own paper.
3. Several of the maps will be used more than once, so students should be told to retain their copies.
4. Some weekly practice series suggest the use of more than one map. In these cases, the students are often expected to gather information from at least two maps and put the information together to produce the answer.

Answer Key for Assessments

page 15: 1. C, 2. B, 3. C, 4. B, 5. D, 6. A

page 16: 1. B, 2. D, 3. C, 4. D, 5. A, 6. D

page 17: 1. C, 2. D, 3. B, 4. C, 5. B, 6. C

page 18: 1. D, 2. B, 3. C, 4. C, 5. A, 6. C

page 19: 1. B, 2. C, 3. A, 4. B, 5. D, 6. C

page 20: 1. D, 2. B, 3. A, 4. D, 5. A, 6. B

page 21: 1. C, 2. A, 3. D, 4. B, 5. D, 6. D

page 22: 1. C, 2. D, 3. B, 4. D, 5. A, 6. C

page 23: 1. B, 2. C, 3. A, 4. C, 5. A, 6. D

page 24: 1. B, 2. D, 3. A, 4. C, 5. C, 6. B

Additional Activities

Unit 1

1. Write a short story about the founding of your community. Do you know when your community was founded? Do you know about the people who founded it? Were there Native American groups in the area? If you do not know exactly, you can make up parts of your story. Pretend you are one of the founders. Why would you choose this place for your community?

2. Is there a special building in your community? Do research to find out about its history. Then, write a short report about the building. Include a drawing of the building with your report.

Unit 2

1. Work in a group of four or five students. Write a short play about some important event in the history of your state. Have each person in the group play a role. Write dialogue for the play. When the play is finished, act it out for the class.

2. Find out about a famous person from your state. Do research to find out more about that person. Then, write a short biography of that person.

3. Work in a group of four or five students. Learn the state song. Sing it for your class.

4. Find a folk tale about your state. With a group of classmates, act out the folk tale in class.

Unit 3

1. Find out about an environmental problem in the region. Make a poster that tells about the problem. Include some pictures. How can the problem be fixed? Display the completed poster in the classroom.

2. To demonstrate the difficulty of mapmaking without the aid of overhead views or sophisticated instruments, have the students make a map of the schoolyard and building. If they are more ambitious, they can make a map of their neighborhood or community. Discuss the difficulties of their mapmaking experience. Display the completed maps in the classroom.

3. Use modeling clay to make a relief map of your region. Show mountains, hills, valleys, plains, rivers, and lakes. Scratch lines into the clay to indicate the state borders. Display your completed map in the classroom.

Unit 4

1. Find the lines of latitude and longitude near your town. Use the lines to estimate your town's location. When you write latitude and longitude, you write the line of latitude first. For example, the latitude and longitude for Milwaukee, Wisconsin, is about 43° N, 88° W. Write the latitude and longitude of your town.

2. Follow the line of latitude near your town east or west. Find a big city somewhere in the world that is near that line. Write the name of the city, its country, and its continent. Now, follow the line of longitude near your town north or south. Find a city somewhere in the world near that line. Write the name of the city, its country, and its continent.

3. Get your teacher or an adult to help you with this activity. Write down the latitude and longtitude for your town. Then, go on the Internet. Go to this address: http://www.fourmilab.ch/cgi-bin/uncgi/Earth. This is the web address for EarthCam. At this site, you can type in your latitude and longitude. Then, you can see your area from a camera in space!

Unit 5

1. Do research to find out about the first explorers to reach the North Pole and the South Pole. Then, write a short report or a short story about one of the explorers who has traveled to the pole.

2. Choose one of the other planets in the solar system. Do research to find out more about that planet. Then, write a report about your planet. Include a drawing of your planet with your report.

3. Write a short story about a trip into space. What would it be like to be the first human visitor to Mars or some other planet?

Name _____ Date _____

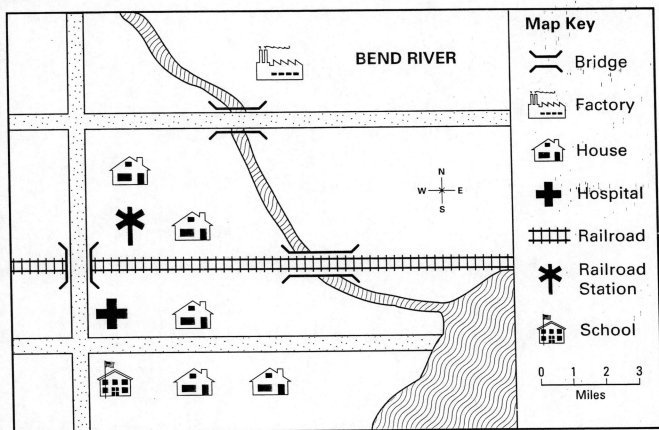

BEND RIVER

Map Key

-)(Bridge
- Factory
- House
- Hospital
- Railroad
- Railroad Station
- School

0 1 2 3
Miles

N
W E
S

Map B

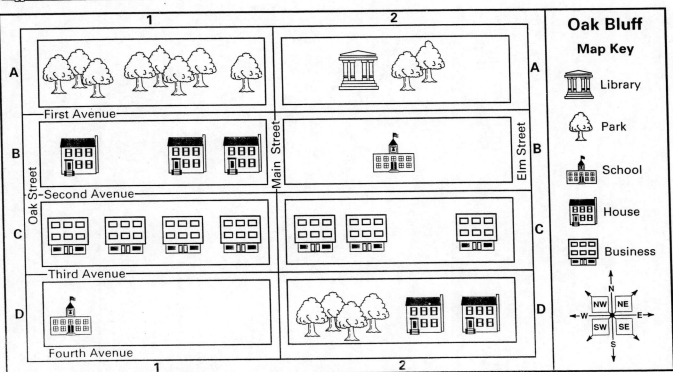

Oak Bluff

Map Key

- Library
- Park
- School
- House
- Business

N
NW NE
W E
SW SE
S

Map C

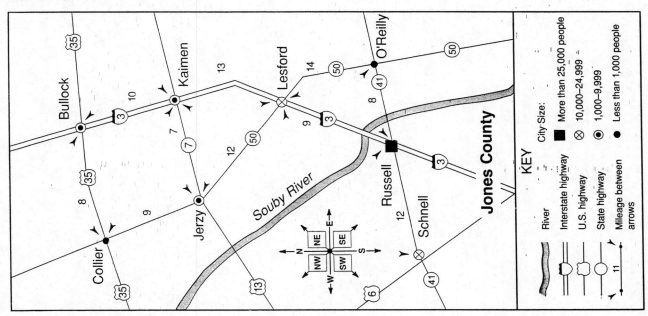

Map D

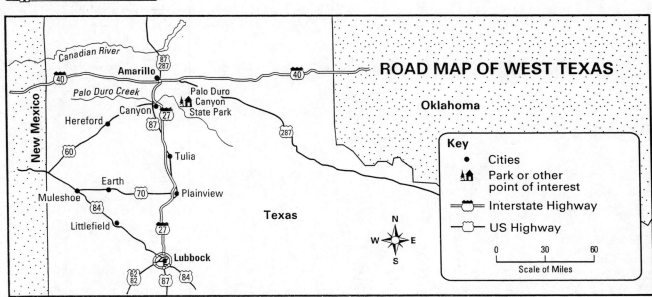

Map E

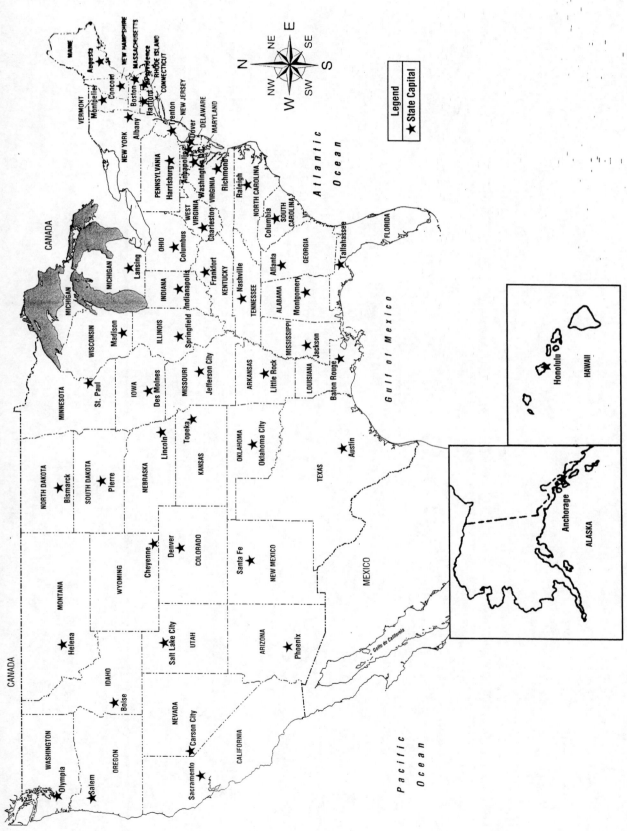

Legend
★ State Capital

Name _____ Date _____

Map F

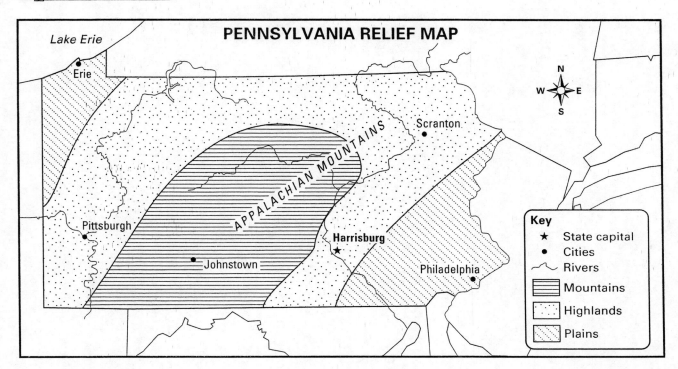

PENNSYLVANIA RELIEF MAP

Lake Erie

Erie

Scranton

APPALACHIAN MOUNTAINS

Pittsburgh

Johnstown

Harrisburg

Philadelphia

Key
★ State capital
● Cities
〜 Rivers
▤ Mountains
▦ Highlands
▨ Plains

Map G

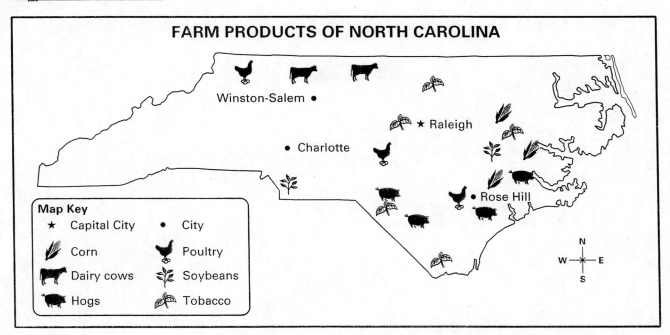

FARM PRODUCTS OF NORTH CAROLINA

Winston-Salem ●

★ Raleigh

● Charlotte

● Rose Hill

Map Key
★ Capital City   ● City
🌽 Corn   🐔 Poultry
🐄 Dairy cows   🌱 Soybeans
🐖 Hogs   🌿 Tobacco

Name _____ Date _____

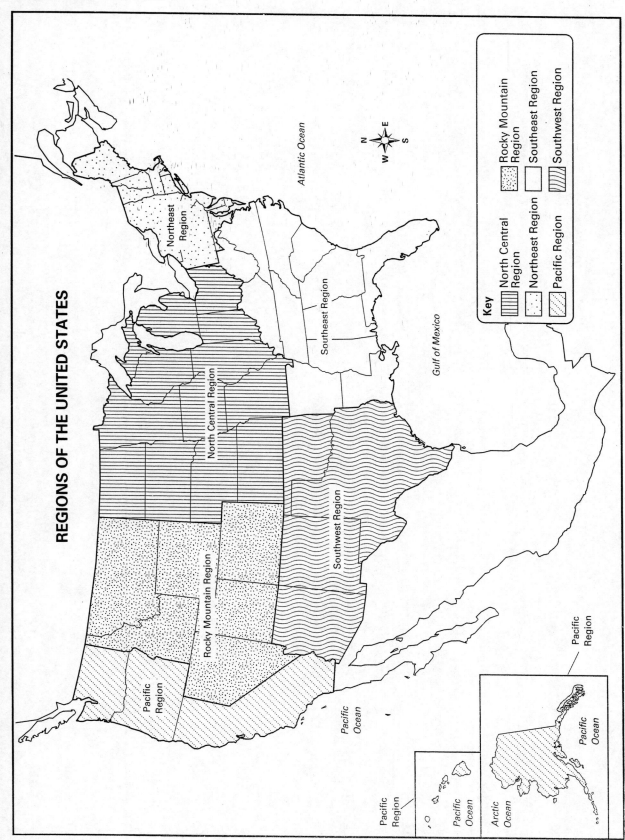

REGIONS OF THE UNITED STATES

Atlantic Ocean

Gulf of Mexico

Northeast Region

Southeast Region

North Central Region

Southwest Region

Rocky Mountain Region

Pacific Region

Pacific Ocean

Pacific Region

Pacific Region

Pacific Ocean

Arctic Ocean

Key

North Central Region

Rocky Mountain Region

Northeast Region

Southeast Region

Pacific Region

Southwest Region

Map I

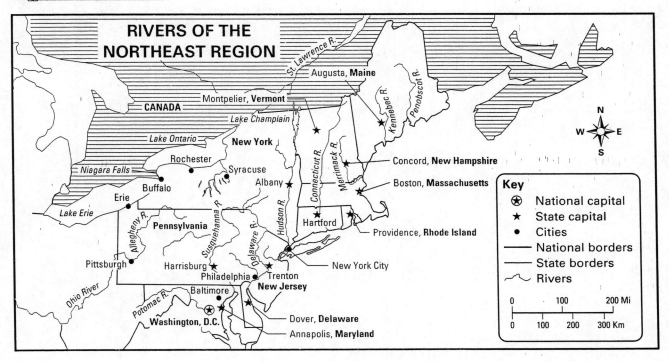

RIVERS OF THE NORTHEAST REGION

CANADA

St. Lawrence R.
Augusta, **Maine**
Kennebec R.
Penobscot R.
Montpelier, **Vermont**
Lake Champlain
Lake Ontario
New York
Merrimack R.
Connecticut R.
Rochester
Syracuse
Albany
Concord, **New Hampshire**
Niagara Falls
Buffalo
Erie
Lake Erie
Hudson R.
Boston, **Massachusetts**
Allegheny R.
Pennsylvania
Susquehanna R.
Delaware R.
Hartford
Providence, **Rhode Island**
Pittsburgh
Harrisburg
New York City
Ohio River
Philadelphia
Trenton
New Jersey
Baltimore
Potomac R.
Washington, D.C.
Dover, **Delaware**
Annapolis, **Maryland**

Key
⊛ National capital
★ State capital
• Cities
▬ National borders
─ State borders
〰 Rivers

0   100   200 Mi
0 100 200 300 Km

Map J

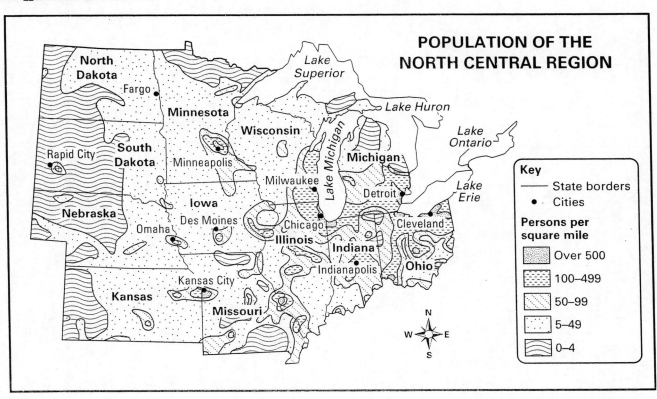

POPULATION OF THE NORTH CENTRAL REGION

North Dakota
Fargo
Lake Superior
Minnesota
Lake Huron
Wisconsin
Lake Ontario
Rapid City
South Dakota
Minneapolis
Lake Michigan
Michigan
Lake Erie
Milwaukee
Detroit
Nebraska
Iowa
Omaha
Des Moines
Chicago
Cleveland
Illinois
Indiana
Ohio
Kansas City
Indianapolis
Kansas
Missouri

Key
─ State borders
• Cities
Persons per square mile
▓ Over 500
⋮ 100–499
╱ 50–99
· 5–49
〰 0–4

Name _____   Date _____

Map K

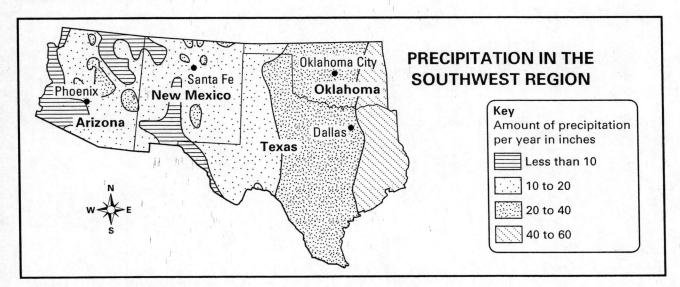

PRECIPITATION IN THE
SOUTHWEST REGION

Oklahoma City
Oklahoma
Santa Fe
New Mexico
Dallas
Phoenix
Arizona
Texas

N
W · E
S

Key
Amount of precipitation
per year in inches

Less than 10
10 to 20
20 to 40
40 to 60

Map L

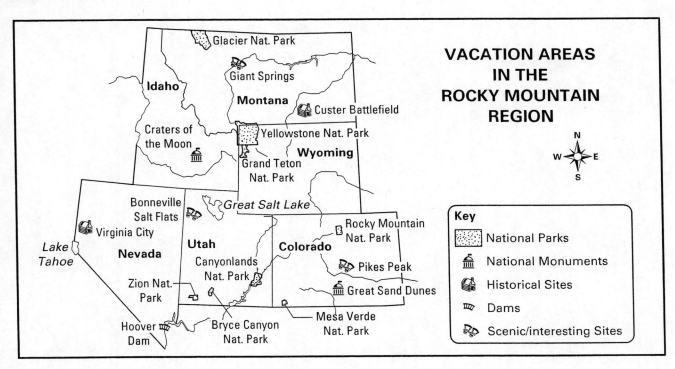

VACATION AREAS
IN THE
ROCKY MOUNTAIN
REGION

N
W · E
S

Glacier Nat. Park
Giant Springs
Idaho
Montana
Custer Battlefield
Craters of
the Moon
Yellowstone Nat. Park
Wyoming
Grand Teton
Nat. Park
Great Salt Lake
Bonneville
Salt Flats
Rocky Mountain
Nat. Park
Virginia City
Lake
Tahoe
Nevada
Utah
Colorado
Canyonlands
Nat. Park
Pikes Peak
Zion Nat.
Park
Great Sand Dunes
Hoover
Dam
Bryce Canyon
Nat. Park
Mesa Verde
Nat. Park

Key
National Parks
National Monuments
Historical Sites
Dams
Scenic/interesting Sites

Name _____ Date _____

Map M

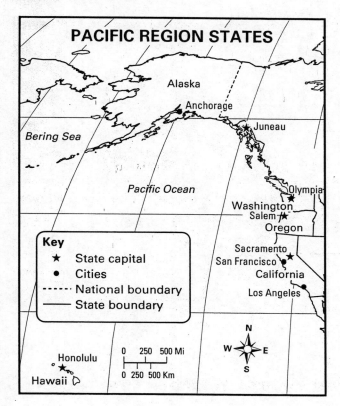

PACIFIC REGION STATES

Alaska
Anchorage
Juneau

Bering Sea

Pacific Ocean

Olympia
Washington
Salem
Oregon

Sacramento
San Francisco
California
Los Angeles

Key
★ State capital
● Cities
---- National boundary
— State boundary

0 250 500 Mi
0 250 500 Km

N
W   E
S

Honolulu
Hawaii

Map N

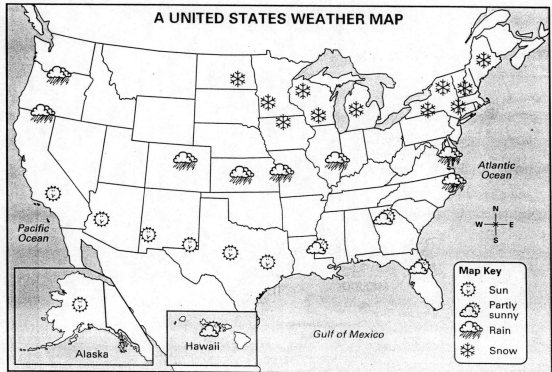

A UNITED STATES WEATHER MAP

Atlantic Ocean

Pacific Ocean

N
W   E
S

Map Key
☼ Sun
⛅ Partly sunny
🌧 Rain
❄ Snow

Alaska
Hawaii
Gulf of Mexico

Name _____ Date _____

Map O

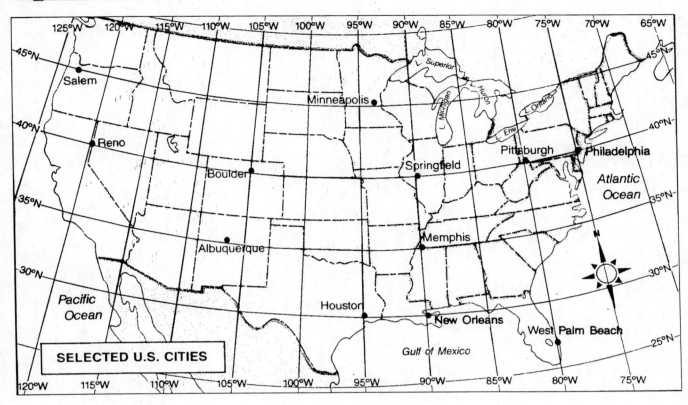

SELECTED U.S. CITIES

Map P

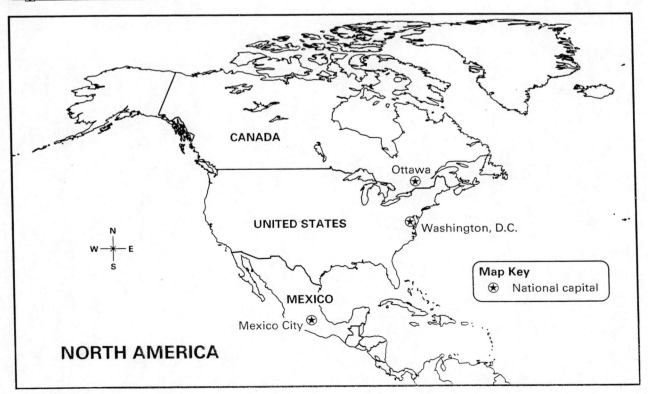

Map Key
⊛ National capital

NORTH AMERICA

Name _____ Date _____

Map Q

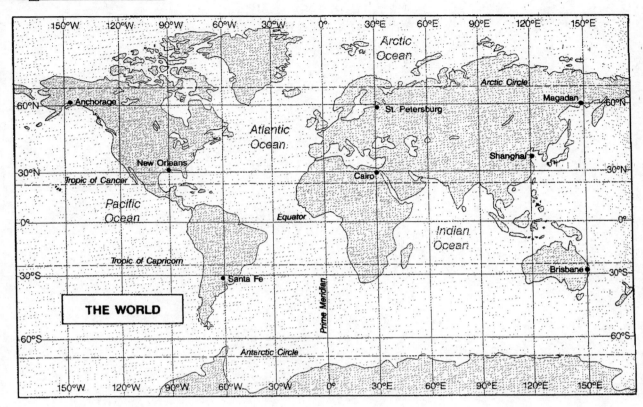

THE WORLD

Map labels: 150°W, 120°W, 90°W, 60°W, 30°W, 0°, 30°E, 60°E, 90°E, 120°E, 150°E

Arctic Ocean, Arctic Circle, Anchorage, St. Petersburg, Magadan, 60°N, Atlantic Ocean, New Orleans, Shanghai, 30°N, Tropic of Cancer, Cairo, Pacific Ocean, Equator, 0°, Indian Ocean, Tropic of Capricorn, Santa Fe, Brisbane, 30°S, Prime Meridian, 60°S, Antarctic Circle

Map R

Mercury Venus Earth Mars Jupiter Saturn Uranus Neptune Pluto

Name _____ Date _____

Map S

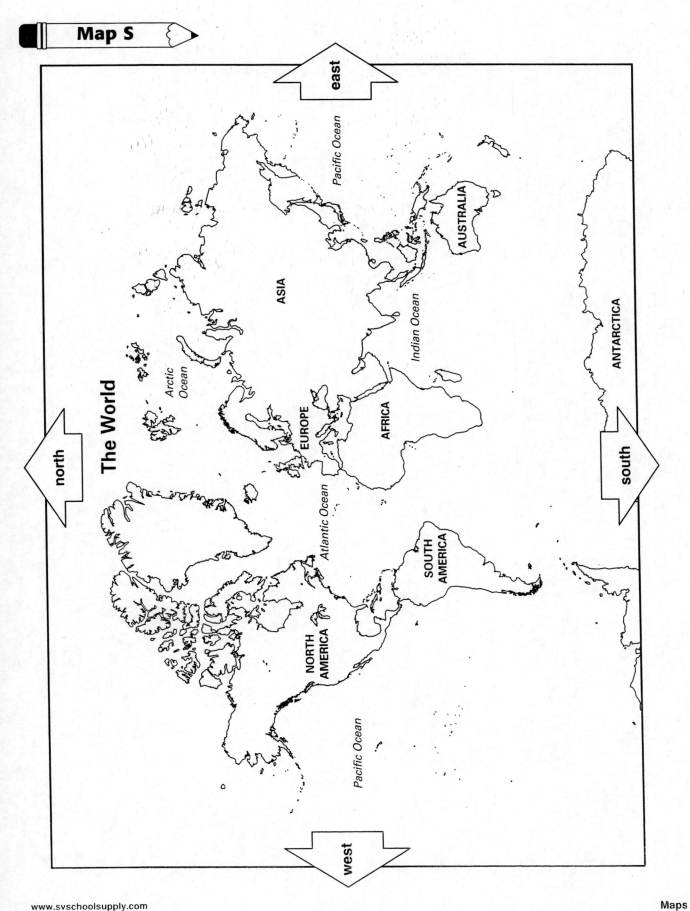

east

The World

north

Arctic Ocean

ASIA

Pacific Ocean

AUSTRALIA

Indian Ocean

EUROPE

AFRICA

Atlantic Ocean

ANTARCTICA

south

NORTH AMERICA

SOUTH AMERICA

Pacific Ocean

west

UNIT 1 PRE-ASSESSMENT

Directions ⟩ Use the map to answer the questions. Darken the circle by the correct answer to each question.

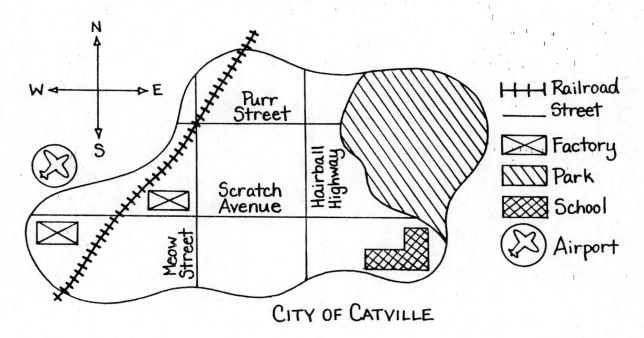

CITY OF CATVILLE

1. What is the name of the city on the map?
- Ⓐ Scratch Avenue
- Ⓑ Airport
- Ⓒ Catville
- Ⓓ Purr Street

2. Which two streets will take you to the park?
- Ⓐ Purr Street and Hairball Highway
- Ⓑ Purr Street and Scratch Avenue
- Ⓒ Meow Street and Scratch Avenue
- Ⓓ Meow Street and Hairball Highway

3. How many factories are in this city?
- Ⓐ none
- Ⓑ one
- Ⓒ two
- Ⓓ three

4. The railroad runs through the intersection of which two streets?
- Ⓐ Purr Street and Hairball Highway
- Ⓑ Meow Street and Purr Street
- Ⓒ Purr Street and Scratch Avenue
- Ⓓ Meow Street and Scratch Avenue

5. In which direction would you travel to get from the park to the airport?
- Ⓐ north
- Ⓑ south
- Ⓒ east
- Ⓓ west

6. In which direction would you travel to get from the school to the park?
- Ⓐ north
- Ⓑ south
- Ⓒ east
- Ⓓ west

Name _____ Date _____

UNIT 1 POST-ASSESSMENT

Directions Use the map to answer the questions. Darken the circle by the correct answer to each question.

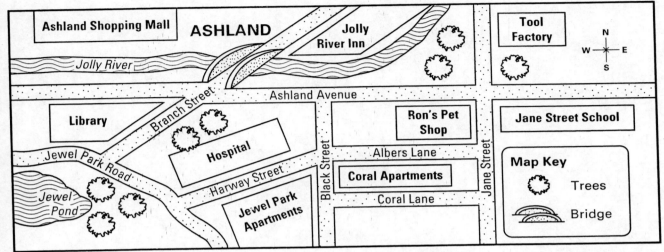

1. What is the name of the city on the map?
- Ⓐ Jewel Park
- Ⓑ Ashland
- Ⓒ Jolly River
- Ⓓ Tool Factory

2. Which street runs by Jewel Pond?
- Ⓐ Black Street
- Ⓑ Coral Avenue
- Ⓒ Ashland Avenue
- Ⓓ Jewel Park Road

3. Which street crosses Jolly River?
- Ⓐ Ashland Avenue
- Ⓑ Jewel Park Road
- Ⓒ Branch Street
- Ⓓ Coral Lane

4. Which direction is the Ashland Shopping Mall from the Jane Street School?
- Ⓐ northeast
- Ⓑ southeast
- Ⓒ southwest
- Ⓓ northwest

5. You live at the Jewel Park Apartments and work at the Tool Factory. Which direction is your job from your home?
- Ⓐ northeast
- Ⓑ southeast
- Ⓒ northwest
- Ⓓ southwest

6. You walk from the Jolly River Inn to the Jewel Park Apartments. Which place would you pass taking Branch Street to Jewel Park Road?
- Ⓐ Jane Street School
- Ⓑ Ron's Pet Shop
- Ⓒ Coral Apartments
- Ⓓ Hospital

Name _____ Date _____

UNIT 2 PRE-ASSESSMENT

Directions Use the map to answer the questions. Darken the circle by the correct answer to each question.

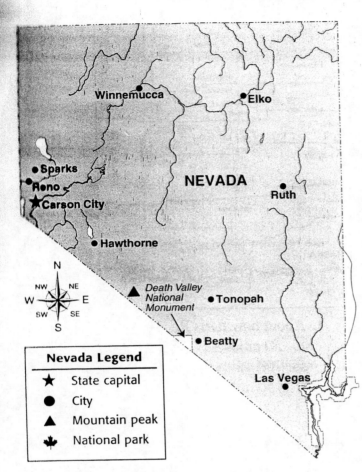

Nevada Legend
★ State capital
● City
▲ Mountain peak
❦ National park

1. What state is shown in the map?
Ⓐ Carson City
Ⓑ Death Valley
Ⓒ Nevada
Ⓓ Legend

2. What is the capital city of this state?
Ⓐ Las Vegas
Ⓑ Ruth
Ⓒ Winnemucca
Ⓓ Carson City

3. What famous place is located near Beatty?
Ⓐ Mount Rushmore
Ⓑ Death Valley National Monument
Ⓒ Yellowstone National Park
Ⓓ Disneyland

4. To travel from Reno to Ruth, which direction must you go?
Ⓐ north
Ⓑ south
Ⓒ east
Ⓓ west

5. To travel from Sparks to Carson City, which direction must you go?
Ⓐ north
Ⓑ south
Ⓒ east
Ⓓ west

6. To travel from Hawthorne to Tonopah, which direction must you go?
Ⓐ northeast
Ⓑ southwest
Ⓒ southeast
Ⓓ southwest

Name _____ Date _____

UNIT 2 POST-ASSESSMENT

Directions ⟩ Use the map to answer the questions. Darken the circle by the correct answer to each question.

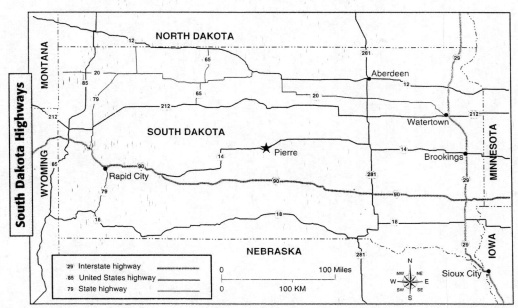

1. Which state borders South Dakota to the north?
- Ⓐ Montana
- Ⓑ Nebraska
- Ⓒ Minnesota
- Ⓓ North Dakota

2. Which state borders South Dakota to the southeast?
- Ⓐ Montana
- Ⓑ Iowa
- Ⓒ North Dakota
- Ⓓ Nebraska

3. If you wanted to fly from Rapid City to Aberdeen, which direction would you go?
- Ⓐ west
- Ⓑ southwest
- Ⓒ northeast
- Ⓓ southeast

4. About how far is Rapid City from Pierre?
- Ⓐ 50 miles
- Ⓑ 100 miles
- Ⓒ 150 miles
- Ⓓ 200 miles

5. Which highway would you use to get from Sioux City, Iowa, to Watertown, South Dakota?
- Ⓐ Interstate Highway 29
- Ⓑ Interstate Highway 90
- Ⓒ United States Highway 18
- Ⓓ United States Highway 212

6. What would be the best route to get from Rapid City to Brookings?
- Ⓐ Interstate Highway 90
- Ⓑ State Highway 79 and United States Highway 18
- Ⓒ Interstate Highway 90 and United States Highway 14
- Ⓓ State Highway 20

Name _____ Date _____

UNIT 3 PRE-ASSESSMENT

Directions Use the map to answer the questions. Darken the circle by the correct answer to each question.

1. Which of these states is in the Southeast Region of the United States?
- Ⓐ Arizona
- Ⓑ Georgia
- Ⓒ Idaho
- Ⓓ Ohio

2. Which of these states is in the Northeast Region of the United States?
- Ⓐ Iowa
- Ⓑ Nevada
- Ⓒ New Jersey
- Ⓓ Arkansas

3. Which of these states is in the Southwest Region of the United States?
- Ⓐ New Mexico
- Ⓑ Oregon
- Ⓒ Wisconsin
- Ⓓ Florida

4. Which of these states is in the North Central Region of the United States?
- Ⓐ Texas
- Ⓑ Minnesota
- Ⓒ South Carolina
- Ⓓ New Jersey

5. Which of these states is in the Pacific Region of the United States?
- Ⓐ Kansas
- Ⓑ Maine
- Ⓒ Tennessee
- Ⓓ California

6. Colorado is in the Rocky Mountain Region of the United States. Which of these states is also in the Rocky Mountain Region?
- Ⓐ Louisiana
- Ⓑ Delaware
- Ⓒ Utah
- Ⓓ Indiana

Name _____ Date _____

UNIT 3 POST-ASSESSMENT

Directions ➤ Use the map to answer the questions. Darken the circle by the correct answer to each question.

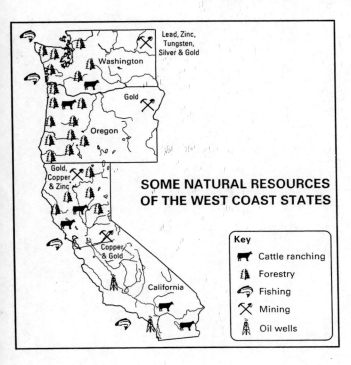

SOME NATURAL RESOURCES
OF THE WEST COAST STATES

Key

🐄 Cattle ranching
🌲 Forestry
🦅 Fishing
⛏ Mining
⛽ Oil wells

1. Which states are shown on this map?
Ⓐ Gold, Copper, and Zinc
Ⓑ Tungsten, Silver, and Gold
Ⓒ Wisconsin, Oregon, and California
Ⓓ California, Washington, and Oregon

2. Where in the United States are these states located?
Ⓐ East Coast
Ⓑ West Coast
Ⓒ Gulf Coast
Ⓓ Gold Coast

3. What part of California uses land for forestry?
Ⓐ north
Ⓑ south
Ⓒ southwest
Ⓓ southeast

4. What mineral is mined in eastern Oregon?
Ⓐ lead
Ⓑ zinc
Ⓒ copper
Ⓓ gold

5. Which state has oil as one of its natural resources?
Ⓐ California
Ⓑ Oregon
Ⓒ Washington
Ⓓ none of these

6. What are two important resources found along the coast of the three states?
Ⓐ mining and oil
Ⓑ fishing and forestry
Ⓒ cattle and mining
Ⓓ none of these

Name _____ Date _____

UNIT 4 PRE-ASSESSMENT

Directions ▷ Use the map to answer the questions. Darken the circle by the correct answer to each question.

1. Which state is directly north of Oregon?
- Ⓐ California
- Ⓑ Idaho
- Ⓒ Washington
- Ⓓ Nevada

2. Which of these states is an island?
- Ⓐ Hawaii
- Ⓑ Florida
- Ⓒ Maine
- Ⓓ Alaska

3. Which of these states is east of Ohio?
- Ⓐ Indiana
- Ⓑ Arkansas
- Ⓒ Wisconsin
- Ⓓ Maryland

4. Which state is directly west of Iowa?
- Ⓐ Illinois
- Ⓑ Nebraska
- Ⓒ Missouri
- Ⓓ Minnesota

5. Which state is directly south of North Dakota?
- Ⓐ Montana
- Ⓑ Wyoming
- Ⓒ South Carolina
- Ⓓ South Dakota

6. Which direction would you travel to get from New York to Wisconsin?
- Ⓐ north
- Ⓑ south
- Ⓒ east
- Ⓓ west

Name _____ Date _____

UNIT 4 POST-ASSESSMENT

Directions ➤ Use the map to answer the questions. Darken the circle by the correct answer to each question.

1. How many states are in the United States?
- (A) 40
- (B) 48
- (C) 50
- (D) 60

2. Which direction would you travel to get from South Dakota to Illinois?
- (A) northeast
- (B) northwest
- (C) southwest
- (D) southeast

3. Which three states touch the southwest corner of Colorado?
- (A) Kansas, Oklahoma, and Texas
- (B) Utah, Arizona, and New Mexico
- (C) Utah, Wyoming, and Idaho
- (D) Wyoming, Nebraska, and Kansas

4. Which of these states shares a border with Mexico?
- (A) Montana
- (B) Louisiana
- (C) Hawaii
- (D) Texas

5. Which of these states touches one of the Great Lakes?
- (A) Ohio
- (B) Iowa
- (C) Utah
- (D) Idaho

6. Which of these states is a peninsula?
- (A) Hawaii
- (B) Missouri
- (C) Florida
- (D) Kentucky

Name _____ Date _____

UNIT 5 PRE-ASSESSMENT

Directions ⟩ Use the map to answer the questions. Darken the circle by the correct answer to each question.

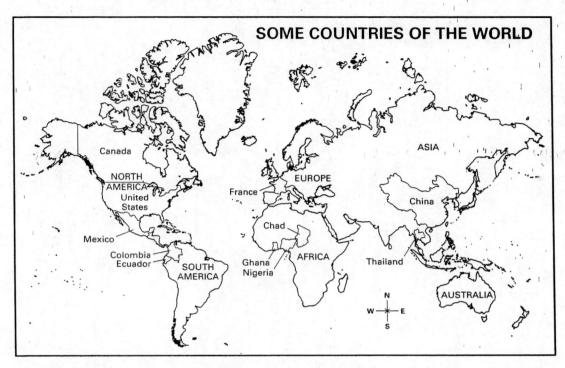

SOME COUNTRIES OF THE WORLD

1. On what continent is the United States located?
- Ⓐ Europe
- Ⓑ North America
- Ⓒ South America
- Ⓓ Africa

2. On what continent is China located?
- Ⓐ Europe
- Ⓑ Africa
- Ⓒ Asia
- Ⓓ Australia

3. On what continent is Nigeria located?
- Ⓐ Africa
- Ⓑ Asia
- Ⓒ Antarctica
- Ⓓ Australia

4. Which continent is an island?
- Ⓐ Europe
- Ⓑ North America
- Ⓒ Australia
- Ⓓ Asia

5. Which continent is directly north of Africa?
- Ⓐ Europe
- Ⓑ Asia
- Ⓒ Antarctica
- Ⓓ Australia

6. To travel directly from Europe to North America, which direction must you go?
- Ⓐ north
- Ⓑ south
- Ⓒ east
- Ⓓ west

Name _____ Date _____

UNIT 5 POST-ASSESSMENT

Directions ⟩ Use the map to answer the questions. Darken the circle by the correct answer to each question.

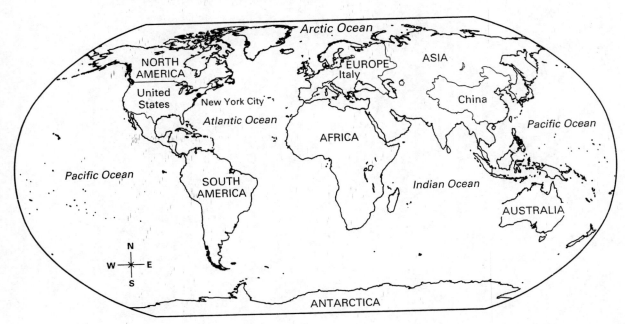

1. To travel directly from Europe to the United States, which ocean must you cross?
- Ⓐ Arctic Ocean
- Ⓑ Atlantic Ocean
- Ⓒ Pacific Ocean
- Ⓓ Indian Ocean

2. To travel directly from Africa to Australia, which ocean must you cross?
- Ⓐ Arctic Ocean
- Ⓑ Atlantic Ocean
- Ⓒ Pacific Ocean
- Ⓓ Indian Ocean

3. Which ocean is the most northern?
- Ⓐ Arctic Ocean
- Ⓑ Atlantic Ocean
- Ⓒ Pacific Ocean
- Ⓓ Indian Ocean

4. Which ocean touches both South America and Asia?
- Ⓐ Arctic Ocean
- Ⓑ Atlantic Ocean
- Ⓒ Pacific Ocean
- Ⓓ Indian Ocean

5. To travel directly from South America to Europe, which direction must you go?
- Ⓐ northwest
- Ⓑ southeast
- Ⓒ northeast
- Ⓓ southwest

6. To travel directly from Europe to Australia, which direction must you go?
- Ⓐ northwest
- Ⓑ southeast
- Ⓒ northeast
- Ⓓ southwest

Assessments

Weekly Geography Practice 4, SV 3433-9

Name _____ Date _____

Monday

1. What is the name of your community?

_____

_____

2. What is your full address and telephone number?

_____

_____

_____

Tuesday

1. Do you live in or near a town, a small city, or a big city?

_____

_____

2. Do you live near an ocean, a lake or river, a forest, a desert, or mountains?

_____

_____

Wednesday

1. About how far do you live from your school?

_____

_____

2. How do you get to school each day?

_____

_____

_____

Thursday

1. What are the fun things to do near your community?

_____

_____

2. Name three things you think make a community a good place to live.

_____

_____

_____

Friday

1. Do you know of a pollution problem in your community? What can people do to stop pollution?

_____

_____

2. Why do you think people chose this place to start your community?

_____

_____

_____

Week Number 1 2 3 4 5 6 7 8 9 10 11 12 13 14 15 16 17 18 19 20 21 22 23 24 25 26 27 28 29 30 31 32 33 34 35 36

Focus: The purpose of this week's questions is to ready the student for some of the concepts associated with geography and maps, such as location, distance, movement, and human interaction with the environment.

1. Answer: Answers will vary.
If you know the origin of the community's name, share it with the students.

2. Answer: Answers will vary.
Review the various parts of an address, such as street name and number, city, state, and ZIP code. See if the students know the area code for their telephone number.

1. Answer: Answers will vary.
Point out to the students that communities come in many different sizes.

2. Answer: Answers will vary.
Point out to the students that communities are founded in many different places, with many differing kinds of surroundings or landforms.

1. Answer: Answers will vary.
Point out to the students that ease of movement is often related to the distance to a place.

2. Answer: Answers will vary.
Review with the students the many different modes of transportation. Perhaps take a poll to see which mode is most used.

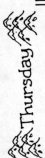

1. Answer: Answers will vary.
Point out to the students that recreational opportunities are an important part of a community.

2. Answer: Answers will vary.
Write all the student responses on the board. Responses might include safety, job opportunities, good schools, good water, recreational opportunities, etc. Take a vote to see which three the students believe are the most important.

1. Answer: Answers will vary.
Clean air and clean water are two important parts of a good quality of life in a community. Point out to the students that people must work together to prevent pollution.

2. Answer: Answers will vary.
Point out to the students that many communities are founded based on a good location. Introduce the concept of the history of the community. Challenge the students to find information about Native American groups that might have lived in the area before the community was settled.

Name _____ Date _____

Monday

1. What can a map show you?

_____

_____

_____

2. What part of a map tells you direction? What are the four cardinal directions?

_____

_____

_____

Tuesday

1. What are the pictures or colors on a map that stand for real things?

_____

_____

2. What is the list of symbols on a map called?

_____

_____

Wednesday

1. What part of a map helps you know how far places really are from each other?

_____

_____

2. What are some ways that people can move from one place to another?

_____

_____

Thursday

1. Why is it good to know about other places?

_____

_____

2. What kinds of things are shown on this map?

_____

_____

Friday

1. What might people do for fun in this town?

_____

_____

2. Why is water an important resource for a community?

_____

_____

Focus: The purpose of this week's questions is to acquaint students with the general features of a map and to make them think beyond the simple physical characteristics of a map.

Directions: Distribute **Map A** to assist students in answering the questions.

1. Answer: where things are located.
Point out that there are many kinds of maps: physical, population, climate and weather, route, resource, cultural. All maps have as their purpose the attempt to help people to understand an area better.

2. Answer: compass rose or direction marker. Answer: north, east, south, west.
Point out that the compass rose allows the map user to know which direction is which. Direction is important in getting from one place to another. You might also note the four intermediate directions: northeast, southeast, southwest, northwest.

1. Answer: symbols.
The symbols are a graphical way of informing the map user where various features are located.

2. Answer: map key or map legend.
The map key is important in allowing the map user to know what all the symbols stand for. Without the map key, the user might not know what the symbols are meant to represent. Then, the map would have little value.

1. Answer: distance scale.
Point out that maps are not full-size. They are scaled. The distance scale allows the user to estimate real distances. Usually the distances are given in miles or kilometers. On **Map A**, one inch on the map equals three miles in real distance.

2. Answer: Answers will vary, but should include the common modes of transportation—by road, by rail, by water, by air, by walking, by using animals.
Point out that people have developed many ways to move from one place to another. Indicate that the physical features of an area may determine the best mode of transportation.

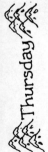

1. Answer: Answers will vary, but should suggest that the knowledge of other places gives one a better understanding of the world in which one lives.
Point out that knowing about other places can help the students to have a better understanding of the place in which they live, simply through comparison and contrast.

2. Answer: **Map A** shows various features such as locations of roads, bridges, railroad tracks, houses, a hospital, a school, and so on.
Point out that the map helps the user to understand how the various features are situated in relation to one another.

1. Answer: The presence of a river and a larger body of water suggests that water activities might be a type of entertainment in this community.
Point out that recreational fun is often determined by the physical features of an area.

2. Answer: Answers will vary, but should suggest that water is necessary for life. People in a community use water to drink, to cook, to clean, to nurture plants and animals, for transportation, for industry, and for recreation.

Name _____ Date _____

🌐 Monday

1. What is the name of this town?

_____

_____

_____

2. What is the symbol for a hospital?

_____

_____

_____

🏙 Tuesday

1. Is the hospital north or south of the school?

_____

_____

2. Is the factory east or west of the river?

_____

_____

🚚 Wednesday

1. How many bridges are in this town?

_____

_____

2. People in this town can travel on the streets or on the railroad. How else can they travel?

_____

_____

⛰ Thursday

1. About how far is it from the railroad station to the school?

_____

_____

2. What is the distance between the two bridges on the river?

_____

_____

🌉 Friday

1. Where could people work in this town?

_____

_____

2. What can people do to take care of this community?

_____

_____

_____

Week Number 1 2 3 4 5 6 7 8 9 10 11 12 13 14 15 16 17 18 19 20 21 22 23 24 25 26 27 28 29 30 31 32 33 34 35 36

Focus: The purpose of this week's questions is to acquaint students with the specific features of a map, especially direction and distance.

Directions: Distribute **Map A** to assist students in answering the questions.

1. Answer: Bend River.
Point out that most maps have titles that tell what the map is about.

2. Answer: a cross.
Point out that students should check the map key to see what the symbols on a map represent.

1. Answer: The hospital is north of the school.
Point out that directions are important to movement.

2. Answer: The factory is east of the river.
Point out that easy movement can be restricted by physical features. Here, people must cross a river bridge to get to the factory.

1. Answer: three bridges.
Point out that the bridges are important to allow easier movement.

2. Answer: They can travel by water or overland by walking or by horseback, for example.
Some students might suggest travel by air, but point out that the map of this community does not show an airport.

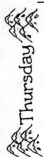

1. Answer: four to five miles.
Point out that knowing the distance between places allows one to calculate travel time better.

2. Answer: five to six miles.
Point out that the river restricts easy movement. People can usually cross the river only at the bridges, so knowing the placement of the bridges and the distance between them allows one to plan travel better.

1. Answer: Three obvious answers are at the school, at the railroad station, and at the factory.
Point out that other people might work as mail carriers or delivery people, as house painters or lawn mowers. There are many less-obvious jobs that the students might suggest.

2. Answer: Answers will vary, but might suggest that the people can keep their community clean and safe, and that they can work to prevent air and water pollution.

Name _____ Date _____

Monday

1. On the map of Oak Bluff, the tree is the symbol for what place?

_____

_____

2. In which grid section is the library located?

_____

_____

Tuesday

1. Which direction is the library from the school in the southwest corner of Oak Bluff?

_____

_____

2. The corner where Elm Street meets Third Avenue is which direction from the houses on Second Avenue?

_____

_____

Wednesday

1. A school is located southeast of the corner where which two streets meet?

_____

_____

2. Which route would you use to get from the library to the school in grid section D-1?

_____

_____

_____

Thursday

1. How many schools are there in Oak Bluff?

_____

_____

2. On which street are all of the businesses located?

_____

_____

_____

Friday

1. Which grid section contains the most park area?

_____

_____

2. Why are parks good for a community?

_____

_____

_____

Focus: The purpose of this week's questions is to acquaint students with the specific features of a map, especially direction, street names, and grid sections.

Directions: Distribute **Map B** to assist students in answering the questions.

1. Answer: park.
Review the use of symbols in the map key to represent places on the map.

2. Answer: A-2.
Indicate to the students that this map is divided into a grid of eight sections. Point out that the grid sections are identified by a letter and number combination, called coordinates, such as A-2. Many map indexes name place locations by using grid coordinates.

1. Answer: northeast.
Point out to the students that they must sometimes be aware of several directional descriptions to locate a place.

2. Answer: southeast.
Point out to the students that directions are often given that refer to the intersection of streets or roads.

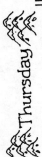

1. Answer: Main Street and First Avenue.
Some students may give the answer Oak Street and Third (or Second) Avenue, which is also not incorrect.

2. Answer: Answers may vary. One route is First Avenue to Oak Street to the school. Another route is First Avenue to Main Street to Fourth Avenue to the school. Accept all possible routes. Point out, though, that the most direct route is often the fastest.

1. Answer: two.
Point out that because the schools are in different locations, directions to each from a given place will be different.

2. Answer: Third Avenue.
Some students might say Second Avenue. Point out that buildings usually have addresses on the street they face. Also note that businesses are often grouped together in the "business section" of a community.

1. Answer: A-1.
Sections A-2 and D-2 also have parkland. Suggest that because A-1 has more tree symbols, it has more parkland.

2. Answer: Answers will vary, but should suggest that parks add greenery and recreational opportunities to a community, as well as promote the quality of life.

Name _____ Date _____

Monday

1. What is the name of the river shown on the map?

_____

_____

2. What is the only interstate highway shown on the map?

_____

_____

Tuesday

1. In what general direction does State Highway 50 run?

_____

_____

2. Which city is located 12 miles east and 9 miles northeast of Schnell?

_____

_____

Wednesday

1. What is the shortest distance from Collier to Kaimen?

_____

_____

2. What would be the shortest route from Bullock to Schnell?

_____

_____

Thursday

1. Which highways are shown crossing the river?

_____

_____

2. What is the largest city in Jones County?

_____

_____

Friday

1. What is the population of O'Reilly?

_____

_____

2. Which city might have the most jobs available?

_____

_____

Focus: The purpose of this week's questions is to acquaint students with the specific features of a map, especially direction, rivers, highways, and city sizes.

Directions: Distribute **Map C** to assist students in answering the questions.

1. Answer: Souby River.
Point out to the students that rivers are usually named on the map and differ in appearance from highways.

2. Answer: Interstate 3.
Point out to the students that there are several classifications of highways, including interstate highways, U.S. highways, and state highways.

1. Answer: southeast to northwest (or northwest to southeast).
Point out to the students that highways often continue in a general direction along their entire length, even though they may twist and curve at times.

2. Answer: Lesford.
Point out to the students that distance on this map is indicated by a number situated between arrows, instead of by a distance scale. Indicate the difference between these mileage indicators and the highway numbers, which are enclosed in some kind of outline.

1. Answer: 16 miles.
Point out that there are at least two possible routes, but one is shorter than the other.

2. Answer: southward along Interstate 3 to Russell, then west on State Highway 41 to Schnell, a distance of 44 miles.
Point out that other routes are possible, but the one above is the shortest.

1. Answer: as shown on the map, Interstate 3, U.S. Highway 13, and State Highway 41 cross the river.
Some of the other highways may cross the river, but they are not shown on the map doing so.

2. Answer: Russell.
Point out to the students the symbols that represent city size in population.

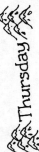

1. Answer: less than 1,000 people.
Point out to the students that people like to live in a place for different reasons. As a result, some cities have larger populations than other cities.

2. Answer: Russell.
Suggest to the students that because Russell has the largest population, it very likely also has the largest number of jobs available.

Directions ⟩ Use **Map E** to help you answer the questions.

Monday

1. The United States is made up of 50 parts. What are these parts called? What is the name of your part?

_____

_____

2. States in the United States are made of many parts. What are these parts called? What is the name of the part you live in?

_____

_____

Tuesday

1. What other states touch the border of your state?

_____

_____

2. Does your state touch the border of another country?

_____

_____

Wednesday

1. Are there any large rivers in your state? How are they used?

_____

_____

2. Is your state near an ocean, a river, a desert, or mountains? How might these places improve or prevent transportation?

_____

_____

Thursday

1. Look at the map. Is your state a big state, a medium-size state, or a small state?

_____

_____

2. Do you have a favorite sports team in your state? What is the team?

_____

_____

Friday

1. Do you know of any pollution problems in your state? How might they be solved?

_____

_____

2. What are some of the important jobs in your state?

_____

_____

Focus: The purpose of this week's questions is to introduce the students to concepts associated with the larger political division of the state.

Directions: Distribute **Map E** to assist students in answering the questions.

1. Answer: states; Answers will vary.
If you know the origin of the state's name, share it with the students. If you know the motto, teach it to the students.

2. Answer: counties; Answers will vary.
If you know the origin of the county's name, share it with the students.

1. Answer: Answers will vary.
Point out that a state often has important links to a neighboring state, such as in economy or regional similarities.

2. Answer: Answers will vary.
Point out that states that share a border with Mexico or Canada often have important links to that neighboring country, such as in economy or immigration.

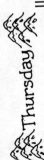

1. Answer: Answers will vary.
Rivers are usually used for transporting people and goods, and for recreational purposes, such as swimming, boating, and fishing. Some rivers are also used to generate electrical power or for agricultural irrigation.

2. Answer: Answers will vary.
Point out that rivers and oceans allow better transportation, and that mountains and deserts usually serve as obstacles to transportation.

1. Answer: Answers will vary.
Point out to the students that states come in many different sizes.

2. Answer: Answers will vary.
Point out that sports teams often have mascots that reflect the local or state history. Sports teams also may serve to promote good-natured regional rivalries.

1. Answer: Answers will vary.
Most states have some sort of pollution problem. Survey the students for innovative solutions to pollution problems.

2. Answer: Answers will vary.
Point out that the healthy economy of a state depends on its people having good jobs.

Name _____ Date _____

Directions → Use **Map E** to help you answer the questions.

Monday

1. What state borders Nebraska to the south?

_____

_____

2. What state borders Alabama to the north?

_____

_____

Tuesday

1. Which state borders only one other state?

_____

_____

2. Which state is farther west—Arizona or New Mexico?

_____

_____

Wednesday

1. In which direction would you travel going from Colorado to Montana?

_____

_____

2. To get directly from Kansas to Illinois, which state would a traveler cross?

_____

_____

Thursday

1. Which two states have only four straight sides?

_____

_____

2. Which state is divided into two large parts, with one part not attached to the other part?

_____

_____

Friday

1. What do we call something found in nature that people can use?

_____

_____

2. What is an important natural resource in your state?

_____

_____

Focus: The purpose of this week's questions is to acquaint the students with various states in the United States and to reinforce the concept of direction.

Directions: Distribute **Map E** to assist students in answering the questions.

1. Answer: Kansas.
You might ask the students to identify states bordering Nebraska in the other cardinal directions.

2. Answer: Tennessee.
You might ask the students to identify states bordering Alabama in the other cardinal directions. Point out that Alabama has a partial border on the Gulf of Mexico.

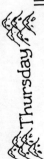

1. Answer: Maine.
Its position in the far northeast corner of the United States limits it to a border only with New Hampshire. Maine, though, does have a border on the Atlantic Ocean, as well as a border with Canada.

2. Answer: Arizona.
Ask the students which state borders Arizona on the west (California).

1. Answer: north.
Review the cardinal directions. Ask the students which direction they would need to travel to get from Montana to Colorado.

2. Answer: Missouri.
Point out that a traveler would also have to cross the Mississippi River to reach Illinois.

1. Answer: Colorado and Wyoming.
Point out that the states have many different shapes. Several states, such as Texas, Oklahoma, Idaho, and Florida, have a shape called a panhandle. See if the students can find these shapes on their map.

2. Answer: Michigan.
The state is divided by Lake Michigan. Point out that the lower half of the state looks like a glove.

1. Answer: natural resource.
Point out some natural resources, such as water, minerals, oil, trees, and seafood.

2. Answer: Answers will vary.
Point out that natural resources are usually very important for a state to have good jobs.

Name _____ Date _____

Directions ⟩ Use **Map E** to help you answer the questions.

Monday

1. What state borders Ohio to the west?

_____

_____

2. What state borders North Dakota to the east?

_____

_____

Tuesday

1. Which state is farther west—Florida or Pennsylvania?

_____

_____

2. Which state is farther north—Nevada or Nebraska?

_____

_____

Wednesday

1. What is a path or road from one place to another called?

_____

_____

2. What is a place where two routes meet called?

_____

_____

Thursday

1. What state borders both Mexico and the Gulf of Mexico?

_____

_____

2. What state or body of water is directly east of Florida?

_____

_____

Friday

1. Why were many communities built at crossroads?

_____

_____

2. Why is it good to know things about other states?

_____

_____

Focus: The purpose of this week's questions is to continue to acquaint students with the various states in the United States and to introduce the concept of routes.

Directions: Distribute **Map E** to assist students in answering the questions.

Monday

1. Answer: Indiana.
You might ask the students to identify states bordering Ohio in the other cardinal directions.

2. Answer: Minnesota.
You might ask the students to identify states bordering North Dakota in the other cardinal directions.

Tuesday

1. Answer: Florida.
Point out that the panhandle of Florida is considerably farther west than the western border of Pennsylvania.

2. Answer: Nebraska.
Point out that the northern border of Nebraska is slightly farther north than the northern border of Nevada.

Wednesday

1. Answer: route.
The general term is *route*, though the students might give answers such as street, highway, trail, way, etc.

2. Answer: crossroads, crossing point, or intersection.
Perhaps you can use an example of a crossroads or intersection near the school.

Thursday

1. Answer: Texas.
Review the different kinds of borders a state might have: another state, another country, a river or lake, an ocean or gulf, etc.

2. Answer: Atlantic Ocean.
Point out how Florida extends into the water. You might want to introduce the term *peninsula*. To the east of Florida is the Atlantic Ocean, to the west is the Gulf of Mexico.

Friday

1. Answer: Crossing points were good stopping places for travelers, who sometimes decided to settle there.
Point out that many of the early routes and trails became the roads and highways of today.

2. Answer: Answers will vary.
Students might suggest that knowing about other states helps them to know more about their country and the different people that live in it.

Name _____ Date _____

Directions > Use **Map D** to help you answer the questions.

Monday

1. Is Amarillo north or south of Lubbock?

_____

_____

2. What cities are directly east of Muleshoe?

_____

_____

Tuesday

1. Which city is farther west—Hereford or Plainview?

_____

_____

2. What major east-west road passes through Amarillo?

_____

_____

Wednesday

1. What river is shown on the map?

_____

_____

2. Name two routes to get from Muleshoe to Canyon.

_____

_____

Thursday

1. What is the name of this map?

_____

_____

2. What is the distance between Amarillo and Lubbock?

_____

_____

Friday

1. What state park is near Amarillo?

_____

_____

2. Why are good roads important for an area?

_____

_____

Focus: The purpose of this week's questions is to acquaint the students with the use of a road map.

Directions: Distribute **Map D** to assist students in answering the questions.

1. Answer: north.
Review the use of the compass rose.

2. Answer: Earth and Plainview.
Point out that many towns have unusual names. Ask the students if they can think of any.

1. Answer: Hereford.
Hereford is more north and more west than Plainview.

2. Answer: Interstate Highway 40.
Most interstate highways that end in an even number run east and west; most interstate highways that end in an odd number run north and south.

1. Answer: Canadian River.
Point out that most roads use a bridge to cross rivers.

2. Answer: U.S. Highway 84 to U.S. Highway 60 to Canyon; U.S. Highway 70 to U.S. Highway 87 (or Interstate Highway 27) to Canyon.
Point out that other routes are possible, but these are the most direct.

1. Answer: Road Map of West Texas.
Review the importance of a title on a map. Titles allow the map user to know what information the map contains.

2. Answer: about 120 miles.
Review the use of a distance scale. On this map, the scale is 60 miles = 1 inch.

1. Answer: Palo Duro Canyon State Park.
Point out that parks make use of the environment to provide recreational opportunities for people.

2. Answer: Good roads make travel easier.

Name _____ Date _____

Directions ▷ Use **Map E** to help you answer the questions.

Monday

1. Which of these cities is the capital of New York: Rochester, Albany, or Buffalo?

_____

_____

2. If you were in Helena, in which state's capital would you be?

_____

_____

Tuesday

1. Which city is the capital of the state west of Georgia and east of Mississippi?

_____

_____

2. Which of these cities is NOT a state capital: Denver, Los Angeles, or Atlanta?

_____

_____

Wednesday

1. Which direction would you travel to get from Santa Fe, New Mexico, to Austin, Texas?

_____

_____

2. Which direction would you travel to get from Raleigh, North Carolina, to Charleston, West Virginia?

_____

_____

Thursday

1. Chicago is the largest city in Illinois, but it is not the capital. Which city is the capital of Illinois?

_____

_____

2. Of which two neighboring states are Bismarck and Pierre the capitals?

_____

_____

Friday

1. Which capital city is named for a large, salty lake?

_____

_____

2. Which capital city is located on an island?

_____

_____

Focus: The purpose of this week's questions is to acquaint the students with various state capitals.

Directions: Distribute **Map E** to assist students in answering the questions.

1. Answer: Albany.

2. Answer: Montana.

1. Answer: Montgomery, Alabama.
Review the importance of knowing directions on a map.

2. Answer: Los Angeles.
You might help students here by pointing out that Los Angeles is in California.

1. Answer: southeast.
Review the use of the compass rose, especially in regard to intermediate directions.

2. Answer: northwest.

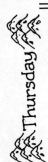

1. Answer: Springfield, Illinois.
Point out that the largest city in a state is often not the capital city.

2. Answer: North Dakota (Bismarck) and South Dakota (Pierre).

1. Answer: Salt Lake City, Utah.
Great Salt Lake is in the northwest part of Utah. Its waters are very salty. In fact, early explorers in the area thought the lake was part of the Pacific Ocean.

2. Answer: Honolulu, Hawaii.
You might define the term *island* as land surrounded on all sides by water. The state of Hawaii is made up of many small islands. Honolulu is on the island called Oahu.

Directions ⟩ Use **Map F** to help you answer the questions.

Monday

1. What city is located on the plains in southeast Pennsylvania?

_____

_____

2. What city is located in the highlands in northeast Pennsylvania?

_____

_____

Tuesday

1. What is the capital city of Pennsylvania?

_____

_____

2. What is the name of this map? What does the map show?

_____

_____

_____

Wednesday

1. Which city is on a lake: Erie, Harrisburg, or Philadelphia?

_____

_____

2. Which city is most likely to have a sports stadium named *Three Rivers Stadium*: Harrisburg, Scranton, or Pittsburgh?

_____

_____

Thursday

1. What is the most common landform in Pennsylvania: mountains, highlands, or plains?

_____

_____

2. Some relief maps label the highest mountain in the state. If this map showed the highest point in Pennsylvania, near which city would you expect it to be: Erie, Johnstown, or Harrisburg?

_____

_____

Friday

1. Several of the cities in Pennsylvania are located beside rivers. Why do you think these cities were started beside rivers?

_____

_____

2. Would you rather live in the mountains or on the plains? Why?

_____

_____

_____

Week Number 1 2 3 4 5 6 7 8 9 10 11 12 13 14 15 16 17 18 19 20 21 22 23 24 25 26 27 28 29 30 31 32 33 34 35 36

Focus: The purpose of this week's questions is to acquaint the students with the use of a relief (or elevation) map.

Directions: Distribute **Map F** to assist students in answering the questions.

1. Answer: Philadelphia.
Point out how the different landforms are represented on the map. You might also point out the relative elevations of each landform: mountains are the tallest, highlands are in the middle, and plains have the lowest elevation.

2. Answer: Scranton.
Review the use of the compass rose and the intermediate directions. Point out that this compass rose does not have *NE*, for example, but it has a small line pointing in that direction.

---

1. Answer: Harrisburg.
Review the use of a map key. Ask the students how the capital city is represented on the map (star).

2. Answer: Pennsylvania Relief Map. It shows the relative heights of the landscape, including mountains, highlands, and plains. It also shows rivers, a lake, cities, and the capital city.

---

1. Answer: Erie.
Ask the students what an important industry in this city might be (shipping).

2. Answer: Pittsburgh.
Point out that Pittsburgh is located where three rivers meet. The Allegheny River and the Monongahela River meet to form the Ohio River.

---

1. Answer: highlands.
Some students might answer mountains; the two have similar proportions. Point out that this fact suggests Pennsylvania has a higher general altitude or elevation.

2. Answer: Johnstown.
Point out that Johnstown is located in the Appalachian Mountains. Mountains have a higher elevation than the surrounding area.

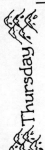

1. Answer: Answers will vary, but should suggest that the rivers allow easier movement of people and goods.

2. Answer: Answers will vary. Conduct a survey of the students and their choices, perhaps including a discussion of their reasons.

Name _____ Date _____

Directions > Use **Map E** to help you answer the questions.

Monday

1. Which state is southeast of Utah?

_____

_____

2. What state borders Tennessee on the northeast?

_____

_____

Tuesday

1. Which states touch the northwest corner of Missouri?

_____

_____

2. Which two capital cities have the state names in their own names?

_____

_____

Wednesday

1. Which state has no common border with any other state or country? (Hint: It's surrounded by water.)

_____

_____

2. Which two states border both Canada and the Atlantic Ocean?

_____

_____

Thursday

1. Which two states each border eight other states, including each other?

_____

_____

2. Which two state names have the name of another state in them?

_____

_____

Friday

1. Where are some good places for recreation, such as camping or swimming, in your state?

_____

_____

2. Name three things that make your state a good place to live.

_____

_____

Focus: The purpose of this week's questions is to acquaint the students with various states in the United States and to reinforce the use of the intermediate directions.

Directions: Distribute **Map E** to assist students in answering the questions.

1. Answer: New Mexico.
Review the intermediate directions: northeast, southeast, southwest, northwest.

2. Answer: Virginia.
Some students might answer North Carolina.

1. Answer: Kansas, Nebraska, and Iowa.

2. Answer: INDIANApolis and OKLAHOMA City.
Many cities have state names in them, but these two are the only capital cities.

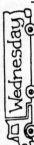

1. Answer: Hawaii.
Review the definition of *island*. Because the Hawaiian Islands touch no other land, they cannot border on any other state.

2. Answer: Maine and New Hampshire.

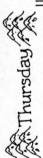

1. Answer: Tennessee and Missouri.
Tennessee borders Missouri, Kentucky, Virginia, North Carolina, Georgia, Alabama, Mississippi, and Arkansas. Missouri borders Tennessee, Arkansas, Oklahoma, Kansas, Nebraska, Iowa, Illinois, and Kentucky.

2. Answer: ArKANSAS and West VIRGINIA.

1. Answer: Answers will vary.
Ask the students if they like to take vacations, and why. What is their favorite vacation spot?

2. Answer: Answers will vary.
Make a list of all the student responses. Take a vote to determine the three most important things.

Week Number 1 2 3 4 5 6 7 8 9 10 11 12 13 14 15 16 17 18 19 20 21 22 23 24 25 26 27 28 29 30 31 32 33 34 35 36

Name _____ Date _____

 **Monday** _____

1. What farm product is grown closest to the capital of North Carolina?

_____

_____

2. What three products are grown or raised near the town of Rose Hill?

_____

_____

_____

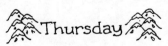 **Tuesday** _____

1. What farm animals are raised in the northern part of the state?

_____

_____

2. What farm animals are raised in the southern part of the state?

_____

_____

Wednesday _____

1. What state or body of water lies directly east of North Carolina? (You can use Map E to help you answer this question.)

_____

_____

2. Why do you think so many farm products are grown in the southeast part of the state?

_____

_____

Thursday _____

1. Are more farm products grown in the east part or the west part of the state?

_____

_____

2. According to the map, what kind of farm animal is raised most in North Carolina?

_____

_____

 **Friday** _____

1. Would you like to live on a farm? Why or why not?

_____

_____

2. What are some important farm products in your state?

_____

_____

Week Number 1 2 3 4 5 6 7 8 9 10 11 12 13 14 15 16 17 18 19 20 21 22 23 24 25 26 27 28 29 30 31 32 33 34 35 36

www.svschoolsupply.com   Unit 2: Sou..
Steck-Vaughn Company   49   Weekly Geography Practice 4...

Focus: The purpose of this week's questions is to acquaint students with the use of a resource map.

Directions: Distribute **Map G** to assist students in answering the questions.

1. Answer: tobacco.
Point out that tobacco is still a main farm crop in many southeastern states, even though tobacco is becoming increasingly unpopular with many people.

2. Answer: poultry, hogs, and corn.
You might need to define *poultry* for the students. Many times, farm crops are grown to feed farm animals, such as corn being used to feed hogs.

1. Answer: dairy cows and poultry.
Point out that dairy cows are used for milk production more than for meat production.

2. Answer: hogs and poultry.
Point out that hogs are raised for meat production. The meat of hogs is called pork, and it includes such cuts as pork chops, bacon, and pigs' feet.

1. Answer: Atlantic Ocean.
Suggest that the students use **Map E** for help in answering this question. North Carolina is a state on the area called the Atlantic Seaboard.

2. Answer: Answers will vary, but should indicate that the closeness of water allows for easier transportation of the farm goods.

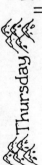

1. Answer: east.
Point out that the east part of the state contains coastal plains. The west part of the state is a part of the Blue Ridge Mountains. Ask the students if plains or mountains are better places to farm.

2. Answer: hogs.
Four hog symbols are included while only three poultry symbols and only two cow symbols are on the map.

1. Answer: Answers will vary.
Ask the students what kind of chores or duties they might have on a farm.

2. Answer: Answers will vary.
Discuss with the students why farms are important.

Name _____   Date _____

Directions ➤ Use **Map H** to help you answer the questions. You may also want to use **Map E**.

Monday

1. What is the name of your region?

_____

_____

_____

2. What other states are in your region?

_____

_____

_____

Tuesday

1. What section of the United States is your region in?

_____

_____

2. What other regions does your region touch?

_____

_____

Wednesday

1. What would be the easiest way to travel from the east side of your region to the west side?

_____

_____

2. How long do you think such a trip would take?

_____

_____

Thursday

1. How is your state similar to other states in your region?

_____

_____

2. How is your state different from other states in your region?

_____

_____

Friday

1. What are some fun vacation spots in your region?

_____

_____

2. Why do you think people like to live in your region? Name three reasons.

_____

_____

Focus: The purpose of this week's questions is to introduce the students to the larger geographical division of the region, including similarities and differences with other states in the same region.

Directions: Distribute **Maps H** and **E** to assist students in answering the questions.

Monday

1. Answer: Answers will vary.
Help students find their state among the regions listed on the map. Students may be able to use **Map E** in conjunction with **Map H** to answer the questions themselves.

2. Answer: Answers will vary.
Help the students to name the other states in the region.

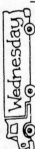
Tuesday

1. Answer: Answers will vary.
Point out that several of the regions have as their name a geographical direction, so this question should be easy for most students.

2. Answer: Answers will vary.
Point out that even though the regions have different names, they are still interrelated, just as the states in a region are interrelated.

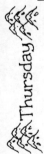
Wednesday

1. Answer: Answers will vary.
Many students will probably answer by car. The *fastest* mode of transportation would be by air.

2. Answer: Answers will vary.
Travel across the Southeast Region, the Southwest Region, the North Central Region, and the Rocky Mountain Region would take several days by car. Travel time across the Pacific Region is harder to determine, because many parts of Alaska are hard to travel across, and Hawaii can only be reached by plane or ship.

Thursday

1. Answer: Answers will vary.
Point out that points of comparison-contrast could include population, climate, natural resources, jobs, landforms, etc.

2. Answer: Answers will vary.

Friday

1. Answer: Answers will vary.

2. Answer: Answers will vary.
Make a list of all the students' responses. Then, take a vote to determine the three most important reasons.

Name _____ Date _____

🌐 Monday

1. In which region is the state of Wisconsin?

_____

_____

2. In which region is the state of Kentucky?

_____

_____

🏙 Tuesday

1. How many regions are in the United States?

_____

_____

2. How many states are in the Rocky Mountain Region?

_____

_____

🚚 Wednesday

1. Could you travel from the Northeast Region to the Southwest Region by water? Through which two bodies of water would you travel?

_____

2. Name two ways you could travel from California to Hawaii in the Pacific Region.

_____

_____

⛰ Thursday

1. What is the smallest state in the Northeast Region?

_____

_____

2. What is the largest state in the Southwest Region?

_____

_____

🌉 Friday

1. In which region would you be more likely to experience a hurricane—the Southeast Region or the Rocky Mountain Region?

_____

_____

2. In which region would you be more likely to experience an earthquake—the Northeast Region or the Pacific Region?

_____

_____

Focus: The purpose of this week's questions is to continue to develop the concept of the geographical region, including similarities and differences between states in different regions.

Directions: Distribute **Maps H** and **E** to assist students in answering the questions.

1. Answer: North Central Region.
You might have students write lists of the states in each region.

2. Answer: Southeast Region.
The Southeast Region and the North Central Region each contain twelve states. They are the two regions with the most states in them.

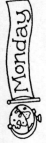

1. Answer: six regions.
Point out that there are several ways to break the United States into regions, and those other divisions might call the regions different names. This map shows six regions in a standard division.

2. Answer: six states.
Pacific Region: five states. Rocky Mountain Region: six states. Southwest Region: four states. North Central Region: twelve states. Southeast Region: twelve states. Northeast Region: eleven states.

1. Answer: Yes; Atlantic Ocean and Gulf of Mexico.
You might point out that people can travel from the Atlantic Ocean to the Pacific Ocean through the Panama Canal.

2. Answer: by airplane or ship.
You might ask the students to list all the different modes of transportation. What is one problem with travel by car (can't cross a large body of water)?

1. Answer: Rhode Island.
Point out the differing sizes of states.

2. Answer: Texas.
You might compare the size of Texas to the size of Alaska. In area, Alaska is over twice as big as Texas.

1. Answer: Southeast Region.
Point out that hurricanes originate in the Atlantic Ocean and the Pacific Ocean. Most that strike the United States hit along the coasts of the Southeast Region or Texas. Hurricanes rarely make it to the Rocky Mountain Region.

2. Answer: Pacific Region.
California, for example, has earthquakes frequently. The Northeast Region very rarely experiences an earthquake.

Name _____ Date _____

Directions > Use **Map I** to help you answer the questions.

Monday

1. Which river flows through Washington, D.C.?

_____

_____

2. Which city is not located at the mouth of a river: Boston, New York, or Philadelphia?

_____

_____

Tuesday

1. Through which city on the map does the Ohio River flow?

_____

_____

2. Name three states whose capital cities are next to rivers.

_____

_____

Wednesday

1. Which river can you use to travel from Trenton, New Jersey, to Philadelphia, Pennsylvania?

_____

_____

2. Which river can you use to travel from Albany, New York, to New York City?

_____

_____

Thursday

1. Which two states have no major rivers flowing through them: Rhode Island and Pennsylvania; Delaware and Rhode Island; or New Hampshire and Vermont ?

_____

_____

2. Which river forms the border between two states and then runs south through two other states: Allegheny River, Connecticut River, or Kennebec River?

_____

_____

Friday

1. Which of these is a human-made feature: a river, a bridge, or a lake?

_____

_____

2. Why do you think many cities of the Northeast Region grew up beside rivers?

_____

_____

Focus: The purpose of this week's questions is to acquaint the students with other features on a physical map, particularly rivers.

Directions: Distribute **Map I** to assist students in answering the questions.

1. Answer: Potomac River.

2. Answer: Boston, Massachusetts.
Though Boston is not on a river, it does have a harbor on the Atlantic Ocean.

1. Answer: Pittsburgh, Pennsylvania.

2. Answer: Students should name three of the following: Pennsylvania, New York, Connecticut, New Hampshire, Maine, New Jersey, Maryland.

1. Answer: Delaware River.
Point out that it is usually easier to travel downstream than upstream on a river. The current flows downstream.

2. Answer: Hudson River.
You might point out the difference between a river's *source* (origin) and its *mouth* (where it feeds into a larger body of water).

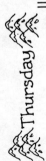

1. Answer: Delaware and Rhode Island.

2. Answer: Connecticut River.
It forms the border between New Hampshire and Vermont, then flows south through Massachusetts and Connecticut.

1. Answer: bridge.
Bridges often have to be built by humans in order to cross rivers.

2. Answer: The rivers were a way to travel from one place to another. There were no highways or railroads. Rivers gave the people a way to get the things they needed to live.

Name _____ Date _____

Monday

1. For which region does this map show information?

_____

_____

2. What state is in the northeast corner of this region?

_____

_____

Tuesday

1. How many people are there per square mile in Chicago: 50–99, 100–499, or over 500?

_____

_____

2. Which of the following cities has between 5 and 49 people per square mile: Fargo, Indianapolis, or Cleveland?

_____

_____

Wednesday

1. Which two large cities are located on Lake Michigan?

_____

_____

2. Which city in Ohio has a port on Lake Erie?

_____

_____

Thursday

1. Which state seems to have the most dense population?

_____

_____

2. Which of these states has fewer people but a larger area of land: South Dakota, Illinois, or Ohio?

_____

_____

Friday

1. Why do you think more people live near the Great Lakes than on the western edge of this region?

_____

_____

2. What can the Great Lakes provide for people who live near them?

_____

_____

Focus: The purpose of this week's questions is to introduce the students to the use of a population map.

Directions: Distribute **Map J** to assist students in answering the questions.

1. Answer: North Central Region.
Review the importance of understanding the title of a map.

2. Answer: North Dakota.

1. Answer: over 500.
Point out the different types of shading used on this map to indicate population density. Chicago is one of the largest cities in the United States.

2. Answer: Fargo, North Dakota.
Point out the kind of shading used in the Fargo area. Indicate to the students that North Dakota has a very low population density.

1. Answer: Chicago, Illinois, and Milwaukee, Wisconsin.
Introduce the term *Great Lakes* and have the students name the five lakes.

2. Answer: Cleveland.

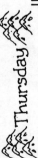

1. Answer: Ohio.
You might define *dense* for the students as meaning most people per square mile. Much of Ohio is shaded for the top three categories of people per square mile.

2. Answer: South Dakota.
South Dakota, like many of the North Central states that do not border the Great Lakes, has a fairly low population density.

1. Answer: The Great Lakes offer job opportunities, especially in shipping. The states away from the Great Lakes tend to have more farming.

2. Answer: The Great Lakes provide transportation, jobs, and recreational opportunities, among other things the students might mention.

Name _____ Date _____

🌐 Monday

1. For which region does this map show information?

_____

_____

2. Which state in this region is farthest south?

_____

_____

🏙 Tuesday

1. What part of Oklahoma receives 40 to 60 inches of rain a year?

_____

_____

2. About how much rain does Phoenix, Arizona, receive each year?

_____

_____

🚚 Wednesday

1. Which direction would you go to travel from Dallas, Texas, to Santa Fe, New Mexico?

_____

_____

2. Which state would probably take the longest to drive across from east to west?

_____

_____

⛰ Thursday

1. According to the map, which state gets more rain per year: Oklahoma or New Mexico?

_____

_____

2. In which state would you expect to see more green plants, grass, and trees: Texas or Arizona?

_____

_____

🌉 Friday

1. How can people in drier states conserve water?

_____

_____

2. How can you use the map to explain why cattle ranching is common in western Texas, while farming is common in eastern Texas?

_____

_____

_____

Focus: The purpose of this week's questions is to introduce the students to the use of a precipitation map.

Directions: Distribute **Map K** to assist students in answering the questions.

1. Answer: Southwest Region.
Have the students identify the four states in the region.

2. Answer: Texas.

1. Answer: eastern part.
Point out the different types of shading used on this map to indicate amounts of precipitation.

2. Answer: about 10 to 20 inches.
Ask the students if they know what kind of land is around Phoenix (desert).

1. Answer: northwest or west.

2. Answer: Texas.
From Beaumont on the eastern border to El Paso on the western tip is about 800 miles by highway.

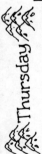

1. Answer: Oklahoma.
Point out that Oklahoma has more areas receiving greater amounts of rainfall. New Mexico, like Arizona, has many desert areas.

2. Answer: Texas.
The eastern half of Texas has large areas that receive the greatest amounts of rainfall, so one could expect to see more greenery there. Arizona has few areas that receive much rainfall.

1. Answer: Answers will vary, but students might suggest: water plants less often; do not water during the heat of the day; wash cars less often; do not leave water running needlessly; install low-flow bathroom fixtures; take fewer baths.

2. Answer: There isn't enough rain in western Texas for farming; crops will not grow. Cattle ranching requires less water than farming.

Name _____ Date _____

Directions Use **Map H** to help you answer the questions. You may also want to use **Map E**.

Monday

1. In which state in the Northeast Region can you find the easternmost city in the United States?

_____

_____

2. Another state besides Kansas in the North Central region is bordered by four states. The first letter of each of these states' names, spelled in a clockwise direction, is I'M OK. What is the state?

_____

_____

Tuesday

1. Which state capitals in the North Central Region are named for Presidents of the United States?

_____

_____

2. What two states in the Pacific Region border both Canada and the Pacific Ocean?

_____

_____

Wednesday

1. If you left Virginia and traveled directly west to California, through which regions would you pass?

_____

_____

_____

2. Which two regions have states that touch the Great Lakes?

_____

_____

Thursday

1. Which of the states in the Southwest Region border Mexico?

_____

_____

2. Which state in the Pacific Region borders Mexico?

_____

_____

Friday

1. Mount St. Helens is a volcano in the Pacific Region. It is located in the state directly north of Oregon. What state is that?

_____

_____

2. How can volcanoes affect people?

_____

_____

_____

Week Number 1 2 3 4 5 6 7 8 9 10 11 12 13 14 15 16 17 18 19 20 21 22 23 24 25 26 27 28 29 30 31 32 33 34 35 36

Focus: The purpose of this week's questions is to introduce the students to the use of a regional map that highlights scenic or recreational areas.

Directions: Distribute **Map L** to assist students in answering the questions.

1. Answer: Wyoming.
Small parts of the park extend into Idaho and Montana.

2. Answer: Colorado.
You might point out that *Colorado* means "colored red" in Spanish.

1. Answer: Bonneville Salt Flats.
Many land-speed record attempts are held at the Bonneville Salt Flats.

2. Answer: Canyonlands National Park, Bryce Canyon National Park, and Zion National Park.
Many national parks are located in the western half of the United States.

1. Answer: north.

2. Answer: Colorado and Wyoming.

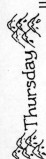

1. Answer: mountains.

2. Answer: a mountain range.

1. Answer: Custer Battlefield. General George Custer's army troops were defeated by many Native American groups.

2. Answer: Nevada. Dams are built to hold back water to create lakes or to prevent flooding downstream. Dams are also used to generate electrical power.

Name _____ Date _____

Monday

1. About how far apart are the Alaskan cities of Anchorage and Juneau?

_____

_____

2. Is Salem, Oregon, east or west of Los Angeles, California?

_____

_____

Tuesday

1. What is the state capital of Washington?

_____

_____

2. If you wanted to travel to the northernmost state in the United States, where would you have to go? What region is this state in?

_____

Wednesday

1. What direction would you travel from Anchorage, Alaska, to Honolulu, Hawaii? About how far apart are the two cities?

_____

_____

2. What is the distance from Los Angeles, California, to Honolulu, Hawaii? Which direction would you have to travel from California to reach Hawaii?

_____

Thursday

1. What kind of landform has water all around it?

_____

_____

2. What is the difference between an island and a peninsula?

_____

_____

Friday

1. What would be good or bad about living on an island?

_____

_____

2. Why do you think mapmakers usually place Hawaii and Alaska in separate boxes next to the other 48 states?

_____

_____

Focus: The purpose of this week's questions is to continue to acquaint the students with regional maps, this time emphasizing distance and direction.

Directions: Distribute **Map M** to assist students in answering the questions.

1. Answer: about 775 miles.
Stress the fact to the students that Alaska is a very large state.

2. Answer: west.

1. Answer: Olympia.

2. Answer: Alaska; the Pacific Region.

1. Answer: southwest; about 3,500 miles.

2. Answer: about 3,000 miles; southwest.
Stress these great distances by pointing out that the distance from New York City to Los Angeles is only about 2,500 miles.

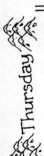

1. Answer: an island.
Ask the students if they know the names of any islands.

2. Answer: An island is land with water all around it, but a peninsula is land not completely surrounded by water.
A peninsula may have water on three sides, such as the strip of land jutting out from the southwest part of Alaska.

1. Answer: Answers will vary.
Some students may say they like water to play in. Other students might note that movement from an island can be restricted.

2. Answer: Answers will vary.
The students should recognize from the map that the distances to Alaska and Hawaii are very large compared to those of the other 48 states. Showing Alaska and Hawaii in separate boxes makes the maps easier to use.

Name _____ Date _____

Monday

1. Which state in the North Central Region is round on the ends and says "hi" in the middle?

_____

_____

2. In which state in the Pacific Region is the Great Salt Lake?

_____

_____

Tuesday

1. Which state capital in the Southeast Region has a name that describes a small stone?

_____

_____

2. Which state in the Pacific Region is the only state named for a United States President?

_____

_____

Wednesday

1. What is the largest kind of waterway: a brook, a stream, or a river?

_____

_____

2. To get directly from Oregon to Florida, through which regions must you pass?

_____

_____

Thursday

1. Which is the only region in which you will find tundra vegetation?

_____

_____

2. Which four states border Lake Erie?

_____

_____

Friday

1. Which state name is also the name of one of the Great Lakes?

_____

_____

2. Why is land next to a waterway or a large lake a good place to start a community?

_____

_____

Focus: The purpose of this week's questions is to continue to develop the concept of the geographical region, including characteristics and landforms.

Directions: Distribute **Maps H, E, or other maps from this unit** to assist students in answering the questions.

Monday

1. Answer: Ohio.

2. Answer: Utah.
Refer the students to **Map L** for help on this question.

Tuesday

1. Answer: Little Rock, Arkansas.

2. Answer: Washington.

Wednesday

1. Answer: river.
Discuss the different sizes of waterways and the way they can contribute to or impede movement.

2. Answer: Rocky Mountain Region, North Central and/or Southwest Region, and Southeast Region.
Depending on the direct line students draw, they may pass through the North Central or the Southwest Region, or through both.

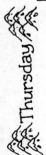

Thursday

1. Answer: Pacific Region; Alaska.
You might define *tundra* as a level or rolling treeless plain that is usually frosty or frozen.

2. Answer: New York, Pennsylvania, Ohio, and Michigan.
Refer the students to **Maps I** and **J** for help on this question.

Friday

1. Answer: Michigan.
Refer the students to **Map J** for help on this question.

2. Answer: Waterways and large lakes can be used to transport people and goods.

Name _____ Date _____

Monday

1. In which country do you live?

_____

_____

2. What is the capital city of your country?

_____

_____

Tuesday

1. Between which two states is Washington, D.C., located?

_____

_____

2. Which state is farther south—California or Florida?

_____

_____

Wednesday

1. Which direction would you travel to get from Louisiana to Idaho?

_____

_____

2. Which direction would you travel to get from South Dakota to South Carolina?

_____

_____

Thursday

1. Which state is located in the middle of the country but borders only four other states?

_____

_____

2. What other countries does your country touch?

_____

_____

Friday

1. Name three reasons your country is a good place to live.

_____

_____

2. If you could pass one law to improve your country, what would your law be?

_____

_____

Focus: The purpose of this week's questions is to acquaint the students with the larger political division of the country, particularly the United States, and some of the basic concepts associated with their country.

Directions: Distribute **Map E** to assist students in answering the questions.

1. Answer: The United States of America.
The assumption is made that this book is being used in the United States.

2. Answer: Washington, D.C.
The assumption is made that this book is being used in the United States.

1. Answer: Maryland and Virginia.
Explain to the students that the city of Washington is not a state, nor is it part of any state. It is a special district: the District of Columbia.

2. Answer: Florida.
You might want to review the cardinal directions.

1. Answer: northwest.
You might want to review the intermediate directions.

2. Answer: southeast.

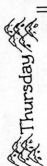

1. Answer: Kansas.
Kansas is roughly the center of the 48 contiguous (touching) states.

2. Answer: Canada and Mexico.
The assumption is made that this book is being used in the United States.

1. Answer: Answers will vary.
Make a list of all the students' responses. Then, take a vote to see which three reasons the students consider the most important.

2. Answer: Answers will vary.
Make a list of all the students' responses. Then, take a vote to see which proposed law the students consider the most reasonable. You might point out that all laws in the United States must abide by the *Constitution*.

Name _____ Date _____

Directions ▷ Use **Map E** to help you answer the questions.

Monday

1. Which state is farther north—Maine or Idaho?

_____

_____

2. The Four Corners is the only place in the United States where you can stand in four states at the same time. Which four states meet at the Four Corners?

_____

_____

Tuesday

1. Which of these is a human-made feature: a plateau, a river, or a highway?

_____

_____

2. Which of these is a physical feature: a bridge, a plain, or a road?

_____

_____

Wednesday

1. Some rivers and streams have shallow places where people can cross by walking, riding, or driving. What are these shallow places called?

_____

2. If a river does not have a shallow place, how can people cross the river?

_____

_____

Thursday

1. If you were flying over a mesa, would you be flying over land or water? What is a mesa?

_____

_____

2. If you were flying over an isthmus, would you be flying over land or water? What is an isthmus?

_____

_____

Friday

1. What kind of landform would be the best place to have a farm: a mountain, a plateau, or a plain?

_____

_____

2. In which part of the United States would you probably find more farmland: the northeastern part or the middle of the country?

_____

_____

Focus: The purpose of this week's questions is to continue to acquaint the students with the concept of country, particularly the United States, and some of the basic landforms associated with their country.

Directions: Distribute **Map E** to assist students in answering the questions.

Monday

1. Answer: Idaho.

2. Answer: Utah, Arizona, New Mexico, and Colorado.
Ask if any students have been to the Four Corners.

Tuesday

1. Answer: a highway.
You might define *plateau* as a mostly level surface raised sharply above the surrounding land on at least one side; a plateau is sometimes called tableland.

2. Answer: a plain.
You might define *plain* as a large area of level or rolling treeless land. Point out that much of the center of the United States is made up of plains, especially the Great Plains.

Wednesday

1. Answer: fords.
Point out that it can be dangerous to cross water in a car or by walking if you do not know how deep the water is.

2. Answer: by bridge, by boat, or by ferry.
Students might suggest other ways; these three are the most common.

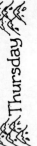

Thursday

1. Answer: land; a flat-topped hill or plateau with steep sides.

2. Answer: land; a narrow strip of land connecting two larger land areas.

Friday

1. Answer: a plain.
Level or rolling plains are the best kind of farmland.

2. Answer: the middle part.
The middle part of the United States contains the Great Plains, which contain good farmland.

Name _____ Date _____

Monday

1. What kind of weather is Hawaii having?

_____

_____

_____

2. What kind of weather is Missouri having?

_____

_____

_____

Tuesday

1. Which two states that touch the Atlantic Ocean are having partly sunny weather?

_____

_____

2. Which two states that touch the Pacific Ocean are having rainy weather?

_____

_____

Wednesday

1. If you drove from Michigan to North Dakota, what kind of weather would you expect to find?

_____

_____

2. If you drove from California to Texas, what kind of weather would you expect to find?

_____

_____

Thursday

1. What kind of weather is the Northeast Region having?

_____

_____

_____

2. What is climate?

_____

_____

_____

Friday

1. Are the snowy parts of the country mostly in the north or in the south?

_____

_____

2. If you wanted to go on vacation where the weather is sunny, which states might you choose to visit?

_____

_____

_____

ocus: The purpose of this week's questions is to acquaint the students with the use of a weather map.

irections: Distribute **Map N** to assist students in answering the questions. You might also suggest
ney refer to **Maps E** and **H** for further assistance.

---

1. Answer: partly sunny.

Point out the different symbols used in the
map key.

2. Answer: rainy.

---

1. Answer: Georgia and Florida.

2. Answer: California, Oregon, or
Washington.

Students should indicate two of these three states.

---

1. Answer: snowy.

Point out that travelers need to pay attention to
the weather so that they do not encounter bad
weather.

2. Answer: sunny.

---

1. Answer: snowy.

Review the locations of the different regions.

2. Answer: the kind of weather a place has
over a long time.

---

1. Answer: north.

2. Answer: California, Arizona, New
Mexico, Texas, or Alaska.

Name _____   Date _____

Monday

1. Which city is the capital of the state that borders Utah on the west?

_____

_____

2. In which of the 50 states can you find the northernmost city in the United States?

_____

_____

Tuesday

1. What state capital in the Southeast Region is named for a President of the United States?

_____

_____

2. The first letters of the names of the Great Lakes spell a word. What is the word?

_____

_____

Wednesday

1. Name four kinds of transportation you could use to get from the east coast to the west coast of the United States.

_____

_____

2. What three very large bodies of water touch the United States on the east, south, and west sides?

_____

_____

Thursday

1. What is the highest kind of landform called?

_____

_____

2. What kind of landform lies between hills or mountains?

_____

_____

Friday

1. In which state would you be more likely to experience a tornado—Rhode Island or Kansas?

_____

_____

2. What kind of job might you have if you lived by the ocean?

_____

_____

Focus: The purpose of this week's questions is to continue to acquaint the students with the concept of country and some of the characteristics and landforms associated with the United States.

Directions: Distribute **Maps E, H, J,** and **M** to assist students in answering the questions.

1. Answer: Carson City, Nevada.

2. Answer: Alaska.

1. Answer: Jackson, Mississippi; Andrew Jackson.

2. Answer: HOMES.
The Great Lakes are Huron, Ontario, Michigan, Erie, and Superior.

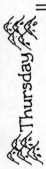

1. Answer: Answers will vary, but could include by air, by rail, by car, by water (around South America or through the Panama Canal), by walking, by horseback, on a bicycle or motorcycle, on a skateboard, etc.

2. Answer: Atlantic Ocean, Gulf of Mexico, and Pacific Ocean.

1. Answer: mountains.
Point out that the United States has two major mountain ranges, the Appalachian Mountains in the east and the Rocky Mountains in the west.

2. Answer: valley.
Some of the more famous valleys in the United States are the Shenandoah Valley in Virginia and Death Valley and Napa Valley in California.

1. Answer: Kansas.
Many more tornadoes strike the plains area than the northeast. The central plains area is sometimes called Tornado Alley. Texas has the greatest frequency of tornado activity.

2. Answer: Answers will vary.
Mostly any kind of job suggestion is possible, but you might point out that an ocean location would also provide jobs in fishing and shipping.

Name _____   Date _____

Monday

1. Lines of latitude run east and west. They are labeled as being so many degrees north or south of the equator. What two cities are located on the line of latitude labeled 30° N?

_____

2. Lines of longitude run north and south. They are labeled as being so many degrees east or west of the Prime Meridian. What two cities are located on the line of longitude labeled 80° W?

_____

Tuesday

1. Locations are indicated by coordinates. Coordinates are numbers that name where lines of latitude and longitude cross. What city is at 40° N, 120° W?

_____

2. What are the coordinates for the city of Springfield? (Hint: give the line of latitude first.)

_____

Wednesday

1. In which direction would you travel going from 40° N latitude and 100° W longitude to 30° N latitude and 100° W longitude?

_____

_____

2. In which direction would you travel going from 40° N latitude and 100° W longitude to 40° N latitude and 90° W longitude?

_____

_____

Thursday

1. What five cities on the map are located on 40° N latitude?

_____

_____

2. If you were following 30° N latitude across the United States, which would be the only three states you would cross?

_____

_____

Friday

1. Would the winter weather be colder at 45° N latitude or 25° N latitude?

_____

_____

2. Would an airplane flying at 30° N, 120° W be over land or water? Why might a pilot want to know the location of the airplane?

_____

_____

Focus: The purpose of this week's questions is to introduce the students to the concepts of longitude and latitude.

Directions: Distribute **Map O** to assist students in answering the questions.

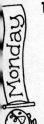

1. Answer: Houston and New Orleans.
Point out to the students that this line of latitude is called 30° N all the way around the Earth. If you have a globe handy, you can demonstrate this fact. Lines of latitude only go up to 90° N or S.

2. Answer: Pittsburgh and West Palm Beach.
Unlike lines of latitude, lines of longitude do not have the same number all the way around the world. Lines of longitude (other than 90°) have a certain number only from pole to pole. Lines of longitude go to 180° E or W. For example, 100° W becomes 80° E on the other side of the world. Use the globe to demonstrate.

1. Answer: Reno.

2. Answer: 40° N, 90° W.
Demonstrate the proper way of giving coordinates.

1. Answer: south.
Demonstrate how to move from one grid intersection to another.

2. Answer: east.

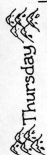

1. Answer: Reno, Boulder, Springfield, Pittsburgh, and Philadelphia.

2. Answer: Florida, Louisiana, and Texas.

1. Answer: 45° N.
The larger the number of the line of latitude, the farther that line is from the equator, so the colder the weather would be.

2. Answer: over the water; Answers will vary.
A pilot would not want to land in the water, for example.

Week Number 1 2 3 4 5 6 7 8 9 10 11 12 13 14 15 16 17 18 19 20 21 22 23 24 25 26 27 28 29 30 31 32 33 34 35 36

Unit 4: Country

Weekly Geography Practice 4, SV 3433-9

Name _____ Date _____

Directions ⟩ Use **Maps E** and **O** to help you answer the questions.

Monday

1. Which state capital in the northeast part of the United States is located east of 70° W longitude?

_____

_____

2. Between which two lines of longitude is Lake Ontario located?

_____

_____

Tuesday

1. Which city is the capital of Wyoming?

_____

_____

2. Which state capital in the northwest part of the United States is located at 45° N latitude?

_____

_____

Wednesday

1. In which direction would you travel going from 35° N latitude and 115° W longitude to 40° N latitude and 105° W longitude?

_____

_____

2. In which direction would you travel going from 45° N latitude and 120° W longitude to 35° N latitude and 110° W longitude?

_____

_____

Thursday

1. Which of these pairs of states borders Tennessee: Indiana and Iowa; West Virginia and Alabama; Missouri and Virginia; Arkansas and South Carolina?

_____

_____

2. Name the states in the United States that have directions in their names.

_____

_____

Friday

1. If you could visit any landform in the United States, such as a mountain or a river, which would you choose to visit? Why?

_____

_____

2. Much of the natural beauty of this country is marred by litter. What can be done to prevent littering?

_____

_____

Focus: The purpose of this week's questions is to continue the study of the United States and the use of latitude and longitude.

Directions: Distribute **Maps E** and **O** to assist students in answering the questions.

1. Answer: Augusta, Maine.

2. Answer: 75° W and 80° W.

1. Answer: Cheyenne, Wyoming.

2. Answer: Salem, Oregon.

1. Answer: northeast.
Demonstrate how to move from one grid intersection to another.

2. Answer: southeast.

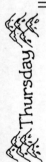

1. Answer: Missouri and Virginia.

2. Answer: North Dakota, South Dakota, West Virginia, North Carolina, and South Carolina.

1. Answer: Answers will vary.
Have the students discuss the reasons for their preferences.

2. Answer: Answers will vary.
Suggested answers might include more public awareness or education about littering, stricter laws, etc.

Name _____ Date _____

Directions ⟩ Use **Map S** to help you answer the questions.

Monday

1. What are the seven largest land areas of the Earth called?

_____

_____

2. On which continent do you live?

_____

_____

Tuesday

1. What are the names of the seven continents?

_____

_____

2. Which continent is the farthest south?

_____

_____

Wednesday

1. If you wanted to travel directly from Europe to Africa, which direction would you go?

_____

_____

2. If you wanted to travel directly from Africa to South America, which direction would you go?

_____

_____

Thursday

1. Which continent is the largest in area?

_____

_____

2. Which continent is the smallest in area?

_____

_____

Friday

1. If you could visit any continent, which would you choose? Why?

_____

_____

2. Many different kinds of people live in the world. Why is it good to know about other kinds of people?

_____

_____

Week Number 1 2 3 4 5 6 7 8 9 10 11 12 13 14 15 16 17 18 19 20 21 22 23 24 25 26 27 28 29 30 31 32 33 34 35 36

Focus: The purpose of this week's questions is to introduce the students to the concept of the world, particularly continents.

Directions: Distribute **Map S** to assist students in answering the questions.

1. Answer: continents.

2. Answer: North America.

1. Answer: North America, South America, Europe, Africa, Asia, Australia, and Antarctica.

2. Answer: Antarctica.

1. Answer: south.
Point out to the students that it is possible to get to Africa from Europe by heading north, but this route is not as direct. Demonstrate with a globe.

2. Answer: west.
Point out to the students that it is possible to get to South America from Africa by heading east, but this route is not as direct. Demonstrate with a globe. Columbus, for example, believed he could reach the East (India) by sailing west from Europe.

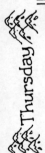

1. Answer: Asia.
Asia has over 17 million square miles in area.

2. Answer: Australia.
Australia has just under 3 million square miles in area.

1. Answer: Answers will vary.
Take a vote to see which continent is the most popular choice.

2. Answer: Answers will vary.
Knowing about other people and other cultures makes people more aware of the different needs and beliefs of their fellow human beings.

Name _____ Date _____

Monday

1. Which three countries make up most of North America?

_____

_____

2. What state of the United States touches Canada's western border? (Hint: you might use **Map M** for this question.)

_____

_____

Tuesday

1. Which city is the capital of Mexico?

_____

_____

2. Which of these cities is the capital of Canada: Ottawa, Montreal, or Toronto?

_____

_____

Wednesday

1. If you wanted to travel from Ottawa to Mexico City, which direction would you go?

_____

_____

2. If you wanted to travel from Mexico City to Alaska, which direction would you go?

_____

_____

Thursday

1. Which country touches the United States on the north?

_____

_____

2. Which country touches the United States on the south?

_____

_____

Friday

1. Much of Canada is cold. Much of Mexico is hot. In which place would you rather live? Why?

_____

_____

2. In far northern places like Alaska, the winter weather is very cold. What would you do if you lived in such a cold climate?

_____

_____

Focus: The purpose of this week's questions is to introduce the students to the continent of North America.

Directions: Distribute **Map P** to assist students in answering the questions.

1. Answer: Canada, the United States, and Mexico.
North America is the third largest of the continents, with about 9.4 million square miles in area.

2. Answer: Alaska.
Point out to the students just how large Alaska is.

1. Answer: Mexico City.
Mexico City is the largest city in population on the North American continent.

2. Answer: Ottawa.

1. Answer: southwest.

2. Answer: northwest.

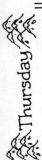

1. Answer: Canada.
Canada is the largest of the three North American countries.

2. Answer: Mexico.
Mexico is the smallest of the three North American countries.

1. Answer: Answers will vary.
Discuss the advantages or disadvantages of living in a hot or cold place.

2. Answer: Answers will vary.
You might also mention the lack of sunshine in northern areas. The sun does not rise above the horizon for many days. Growing seasons are usually quite short.

Name _____ Date _____

Monday

1. What are the largest bodies of water on the Earth called?

_____

_____

2. What are the names of the four oceans?

_____

_____

_____

_____

Tuesday

1. Which ocean touches the United States on the west?

_____

_____

2. Which ocean touches the United States on the east?

_____

_____

Wednesday

1. If you wanted to sail directly from Africa to Australia, which ocean would you cross?

_____

_____

2. If you wanted to sail directly from North America to Africa, which ocean would you cross?

_____

_____

Thursday

1. Which places are the farthest north and the farthest south you can go on the Earth?

_____

_____

2. Which ocean is near the North Pole?

_____

_____

Friday

1. Would you want to swim in the ocean near Antarctica? Why or why not?

_____

_____

_____

2. Why are the oceans important to people?

_____

_____

_____

Focus: The purpose of this week's questions is to continue to acquaint the students with the concept of the world, particularly oceans.

Directions: Distribute **Map S** to assist students in answering the questions.

1. Answer: oceans.
Point out that there are many kinds of bodies of water: oceans, seas, gulfs, bays, lakes, ponds, etc. Oceans are the largest of these.

2. Answer: Atlantic Ocean, Pacific Ocean, Arctic Ocean, and Indian Ocean.
The Pacific Ocean is the largest, covering about 70 million square miles. It is over twice as large as the Atlantic Ocean, which is the second largest.

1. Answer: Pacific Ocean.

2. Answer: Atlantic Ocean.

1. Answer: Indian Ocean.
The Indian Ocean is the third largest ocean, just a bit smaller than the Atlantic Ocean.

2. Answer: Atlantic Ocean.

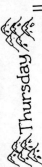

1. Answer: North Pole and South Pole.
Demonstrate the location of the two poles on a globe.

2. Answer: Arctic Ocean.
Use a globe to show students where the North Pole is. The Arctic Ocean and North Atlantic Ocean are very cold, usually with icebergs. The *Titanic* struck an iceberg in the North Atlantic Ocean.

1. Answer: Answers will vary, but should probably indicate the students would not want to swim there because the water is very cold.

2. Answer: Answers will vary.
The students should be aware of the oceans' role in weather and the water cycle, their role in supplying food, and their role in transportation and recreation.

Name _____   Date _____

Monday

1. What is the name of the imaginary line around the Earth halfway between the North Pole and the South Pole?

_____

_____

2. Which directions does this line run? Is it a line of latitude or longitude? What is its number label?

_____

_____

Tuesday

1. What is the name of the imaginary line that runs north and south between the North Pole and the South Pole at 0°?

_____

_____

2. Near which continent do these two imaginary lines meet? Is the meeting point on land or on water?

_____

_____

Wednesday

1. Which two cities are located on latitude line 30° N? Which one is located at 30° N, 30° E?

_____

_____

2. Which direction would you travel to get directly from Brisbane to Cairo? Would you cross the Equator? Would you cross the Prime Meridian?

_____

_____

Thursday

1. Which continent is located completely between 30° W and 90° W?

_____

_____

2. What city is located at 60° N, 150° E? On what continent is this city?

_____

_____

Friday

1. What if you could skate around the world on latitude 30° N? What do you think you would see along the way?

_____

_____

2. Write a short story about your skating trip around the world. Include some pictures with your story.

Focus: The purpose of this week's questions is to reinforce the concepts of latitude and longitude, this time in relation to locations across the world.

Directions: Distribute **Map Q** to assist students in answering the questions.

1. Answer: Equator.
Equator comes from the Latin and means "equalizer."

2. Answer: east and west; latitude; 0°.

1. Answer: Prime Meridian.
The Prime Meridian runs through the site of the Royal Observatory at Greenwich, England. *Meridian* is from Middle English, meaning "at noon."

2. Answer: Africa; on water.
Point out the location to the students on a map or globe.

1. Answer: New Orleans and Cairo; Cairo.
Have the students run their finger along this line of latitude.

2. Answer: northwest; Equator: yes; Prime Meridian: no.

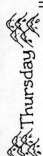

1. Answer: South America.

2. Answer: Magadan; Asia.

1. Answer: Answers will vary.
Have the students volunteer their ideas. Have them check to see which countries are located along the line.

2. Answer: Answers will vary.
This question is intended as a writing extension, and it should be assigned as homework. The students can make up a little book with their story and drawings.

Name _____ Date _____

Directions → Use **Maps Q** and **S** to help you answer the questions.

Monday

1. On which continent is the city of Brisbane located?

_____

_____

2. On which continent is the city of Cairo located?

_____

_____

Tuesday

1. Which ocean is west of Africa and east of South America?

_____

_____

2. Which two oceans touch both North America and Asia?

_____

_____

Wednesday

1. If you wanted to travel from the east coast of South America to the west coast of North America, which direction would you go?

_____

_____

2. If you wanted to travel directly from the south coast of Asia to the north coast of Antarctica, which direction would you go?

_____

_____

Thursday

1. What is a half of the Earth called?

_____

_____

2. In which two hemispheres do you live?

_____

_____

Friday

1. Pollution is a worldwide problem. What can you do to help stop water and air pollution?

_____

_____

2. Earth Day is held each April. Earth Day celebrates the Earth. Make an Earth Day poster with a clever slogan to help stop pollution.

Week Number 1 2 3 4 5 6 7 8 9 10 11 12 13 14 15 16 17 18 19 20 21 22 23 24 25 26 27 28 29 30 31 32 33 34 35 36

Focus: The purpose of this week's questions is to continue to acquaint the students with the concept of the world, and to introduce the concept of hemisphere.

Directions: Distribute **Maps Q** and **S** to assist students in answering the questions.

Monday

1. Answer: Australia.

Point out to the students that Australia is the only continent that contains only one country. The continent and the country have the same name.

2. Answer: Africa.

Tuesday

1. Answer: Atlantic Ocean.

2. Answer: Pacific Ocean and Arctic Ocean.

Wednesday

1. Answer: northwest.

2. Answer: south to southwest.

Thursday

1. Answer: hemisphere.

Point out to the students that the Equator divides the Earth into the Northern Hemisphere and the Southern Hemisphere. The Prime Meridian divides the Earth into the Eastern Hemisphere and the Western Hemisphere.

2. Answer: North America is in the Northern Hemisphere and the Western Hemisphere.

Friday

1. Answer: Answers will vary.

Discuss the students' answers. Possible responses include disposing of chemicals properly, conserving water, using fewer fossil fuels, etc.

2. Answer: Answers will vary.

This is best assigned as homework. Display the completed posters in the classroom.

Name _____ Date _____

Monday

1. What is a globe?

_____

_____

_____

2. What shape is the globe? What shape is the Earth?

_____

_____

Tuesday

1. What is located at the very top of the globe?

_____

_____

2. What is located at the very bottom of the globe?

_____

_____

Wednesday

1. Could you reach Africa from South America by traveling west?

_____

_____

2. Could you reach Asia from Australia by traveling south?

_____

_____

Thursday

1. In which two hemispheres is the continent of Australia?

_____

_____

2. In how many hemispheres is the continent of Africa?

_____

_____

Friday

1. People live all over the world. They live in very cold places and in very hot places. They live in very dry places and very wet places. Why do you think people choose to live where they do?

_____

2. If you could live anywhere in the world, where would you want to live? Why?

_____

_____

_____

Week Number 1 2 3 4 5 6 7 8 9 10 11 12 13 14 15 16 17 18 19 20 21 22 23 24 25 26 27 28 29 30 31 32 33 34 35 36

Focus: The purpose of this week's questions is to introduce the students to the use of a globe to locate places.

Directions: Have a globe available to assist students in answering the questions.

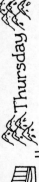

1. Answer: a model of the Earth.

2. Answer: round, or more correctly, spherical; round or spherical.

---

1. Answer: North Pole.
You might point out that from the North Pole, all directions are south. Similarly, from the South Pole, all directions are north.

2. Answer: South Pole.
You might point out that there is a multinational research facility at the South Pole.

---

1. Answer: yes.
You might demonstrate this fact by following a line of latitude around the world.

2. Answer: yes.
You might demonstrate this fact by following a line of longitude over the pole.

---

1. Answer: Eastern Hemisphere and Southern Hemisphere.

2. Answer: all four hemispheres.

1. Answer: Answers will vary.
You might lead a discussion about why people live where they do. Some do so because their ancestors lived there; others move to a place to get a job; others do not move because they lack means of transportation.

2. Answer: Answers will vary.
You might send this question home as a homework assignment.

Name _____   Date _____

Directions > Use **Maps Q** and **S** to help you answer the questions.

Monday

1. On which continent is the city of Anchorage located?

_____

_____

2. On which continent is the city of Shanghai located?

_____

_____

Tuesday

1. What city is located at 60° N, 30° E?

_____

_____

2. What city is located at 30° S, 60° W?

_____

_____

Wednesday

1. Which two directions could you go to travel from Cairo to New Orleans?

_____

_____

2. Which two directions could you go to travel from Magadan to Brisbane?

_____

_____

Thursday

1. Is Antarctica in the Northern Hemisphere or Southern Hemisphere?

_____

_____

2. Is South America in the Eastern Hemisphere or Western Hemisphere?

_____

_____

Friday

1. Why is the study of geography important?

_____

_____

2. Would you like to be a mapmaker?

_____

_____

Focus: The purpose of this week's questions is to continue to acquaint the students with the concepts of world and hemisphere, and with the use of latitude and longitude to locate places.

Directions: Distribute **Maps Q** and **S** to assist students in answering the questions.

Monday

1. Answer: North America.

2. Answer: Asia.

Tuesday

1. Answer: St. Petersburg.
Review the use of coordinates to locate a place.

2. Answer: Santa Fe.

Wednesday

1. Answer: east or west.
Remind the students that the world is spherical.

2. Answer: north or south.

Thursday

1. Answer: Southern Hemisphere.

2. Answer: Western Hemisphere.

Friday

1. Answer: The study of geography helps us to know more about the world we live in.

2. Answer: Answers will vary.
Point out that cartographers, or mapmakers, now use satellites and computers to help them prepare maps.

Week Number 1 2 3 4 5 6 7 8 9 10 11 12 13 14 15 16 17 18 19 20 21 22 23 24 25 26 27 28 29 30 31 32 33 34 35 36

Name _____ Date _____

Monday

1. What are the large bodies called that orbit around the Sun?

_____

_____

2. What is the name of the planet you live on?

_____

_____

Tuesday

1. Which planet is closest to the Sun?

_____

_____

2. Which planet is farthest from the Sun?

_____

_____

Wednesday

1. If you traveled from Earth to Neptune, would you be traveling toward the Sun or away from the Sun?

_____

_____

2. How long do you think it would take to travel to Neptune from Earth?

_____

_____

Thursday

1. Which is the largest planet?

_____

_____

2. Which planet has many rings?

_____

_____

Friday

1. Why can't people live on Mercury?

_____

_____

_____

2. Would you like to travel into space? Why?

_____

_____

_____

Focus: The purpose of this week's questions is to acquaint the students with the concept of the solar system and Earth's place in it.

Directions: Distribute **Map R** to assist students in answering the questions.

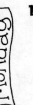

Monday

1. Answer: planets.
The word *planet* comes from the Greek and Latin, meaning "wanderer."

2. Answer: Earth.

Tuesday

1. Answer: Mercury.

2. Answer: usually Pluto.
At times, Pluto moves inside the orbit of Neptune.

Wednesday

1. Answer: away from the Sun.

2. Answer: many, many years.
Neptune is over 4 billion kilometers (about 2.6 billion miles) from Earth.

Thursday

1. Answer: Jupiter.

2. Answer: Saturn.
Saturn is the best-known planet with rings, though several of the outer planets have rings. Saturn's rings are the most prominent.

Friday

1. Answer: It is too hot on the side toward the Sun and too cold on the side away from the Sun.
Earth is the only planet known to have a moderate climate.

2. Answer: Answers will vary.
